SPACE CONQUEST

THE COMPLETE HISTORY OF MANNED SPACEFLIGHT

FRANCIS DREER

TRANSLATED BY KEN SMITH

Haynes Publishing

First published as *Conquête Spatiale: Histoire Des Vol Habités*, in 2007 in France by E-T-A-I, 20 rue de la Saussiere, 92641 Boulogne-Billancourt Cedex

This English-language edition published in January 2009 by Haynes Publishing

A catalogue record for this book is available from the British Library

ISBN 978 1 84425 573 3

Library of Congress catalog control no. 2008939591

Published by Haynes Publishing,
Sparkford, Yeovil, Somerset BA22 7JJ, UK
Tel: 01963 442030 Fax: 01963 440001
Int. tel: +44 1963 442030
Int. fax: +44 1963 440001
E-mail: sales@haynes.co.uk
Website: www.haynes.co.uk

Haynes North America Inc.,
861 Lawrence Drive, Newbury Park, California 91320, USA

Printed and bound in Britain by J. H. Haynes & Co. Ltd, Sparkford

CONTENTS

Glossary of acronyms

AAP Apollo Application Programme
ACES Advanced Crew Escape Suit
ALSEP Apollo Lunar Surface Experiments Package
ALT Approach and Landing Test
AM Adaptor Module
AMU Astronaut Manoeuvring Unit
AS Apollo-Saturn
ASTP Apollo-Soyuz Test Project
ATDA Augmented Target Docking Adaptor
ATM Apollo Telescope Mount
ATV Agena Target Vehicle
CDR Commander
CEV Crew Exploration Vehicle
CLV Crew Launch Vehicle
CMP Command Module Pilot
COSTAR Corrective Optics Space Telescope Axial Replacement
CSM Command and Service Module
DA *Dal'naya Aviatsiya*, 'long-range flight'
DFI Development Flight Instrumentation
DPS Descent Propulsion System
EASEP Early Apollo Scientific Experiment Package
ECS Environmental Control System
EM Equipment Module
EMU Extra-vehicular Mobility Unit
EOR Earth Orbit Rendezvous
ESA European Space Agency
ESP External Storage Platform
ET External Tank
EVA Extra-Vehicular Activity
FCB (or FGB) Functional Cargo Block
ICBM Inter-Continental Ballistic Missile
ICC Integrated Cargo Carrier
ICM Interim Control Module
ILRV Integrated Launch and Re-entry Vehicle
IPS Instrument Pointing System
ISS International Space Station
ITC Integrated Truss Structure
JPL Jet Propulsion Laboratory
KHSC Krunichev Space Research Centre
LCG Liquid Cooling Garment
LEM Lunar Exploration Module
LLRV Lunar Landing Research Vehicle
LM Lunar Module
LMTV Lunar Module Training Vehicle
LOR Lunar Orbit Rendezvous
LRS Launch Escape System
LWT Lightweight Tank
MATS Military Air Transport Service
MET Modular Equipment Transporter
MKBS Multi-Module Space Station
MMU Manned Manoeuvring Unit
MOL Manned Orbiting Laboratory
MRBM Medium Range Ballistic Missile
MSS Mobile Servicing System
MT Mobile Transporter
NACA National Advisory Committee for Aeronautics
NASA National Aeronautics and Space Administration
OAMS Orbit Attitude and Manoeuvring System
OBSS Orbiter Boom Sensor System
OKB *Opytnoe Konstrucktorskoe Byuro*
OWS Orbital Work Shop
PLSS Personal Life Support System
PTC Passive Thermal Control
RM Re-entry Module
RMS Remote Manipulator System
SA Saturn-Apollo
SAC Strategic Air Command
SCA Shuttle Carrier Aircraft
SIM Scientific Instrument Module
SLWT Super-Lightweight Tank
SPAS Shuttle Pallet Satellite
SPS Service Propulsion System
SRB Solid Rocket Booster
SSME Space Shuttle Main Engine
TDRS Tracking and Data Relay Satellite
TEI Trans-Earth Injection
TISP Teacher In Space Programme
TLI Trans-Lunar Injection
USAF United States Air Force
VAB Vertical Assembly Building

Photo credits

NASA (NASA-GRIN) (NASA-JSC)
Cover and pages 10–15, 18 bottom, 25–30, 32–5, 37, 40, 43, 44 centre, 46–8, 49 top right and bottom, 50–3, 55 top and bottom, 56–72, 73 bottom left, 74–6, 78–85, 88–103, 105–15, 118–25, 128–9, 130 bottom, 131–5, 138 top and centre, 139–40, 141 top, 145–57, 158, 159 right, 162–7, 170, 174–83, 186, 187–9, 190 right, 191–202, 203 bottom, 204–7.

NASA/Scott Andrews
Page 203 top.

DR/RKA
Pages 16, 19 top, 20 bottom, 21 top, 41–2, 43 second down, 44 top and bottom, 54 bottom, 73 top, 86, 87 top, 117, 143 bottom, 144, 169, 171–3, 184–5.

DR/big-block.com collection
Pages 17, 18 top, 20 top, 21 bottom, 22, 24 bottom, 31, 36, 44–5, 49 top left, 55 centre, 73 bottom right, 87 bottom, 104 top, 117 top, 138 bottom, 141 bottom, 142, 143 top, 159 centre, 163 centre.

Mark Wade/Astronautix.com
Page 19 bottom.

Everything has been done to locate copyright holders. We apologise in advance for any unintended errors or omissions, which we would be happy to correct in any future edition. We hope that this edition will reveal the names of any copyright holders, to whom we reserve the usual rights.

Acknowledgements

NASA
www.nasa.gov

NASA-GRIN
http://grin.hq.nasa.gov

Mark Wade
www.astronautix.com

Didier Capdevila
www.capcomespace.net

Russian Space Agency
http://liftoff.msfc.nasa.gov/rsa/rsa.html

Energia
http://www.energia.ru/english/

Anatoly Zak
http://www.russianspaceweb.com/

Asif Siddiqi-Soviet Web Space
http://home.earthlink.net/-cliched/main_space.html
(Publisher's note: this website no longer existed at the time of publication)

PREFACE

Half a century ago, a sphere less than a metre in diameter was sent into orbit by the Soviet Union. Sputnik 1, as this first satellite was called, had just started what would prove to be a frantic dash for the stars. The conquest of space has remained a peaceful affair, but looking back it is very clear that a battle was being fought between the two superpowers of the time, the United States of America and the Soviet Union. Yet in 1961, when the young President John Fitzgerald Kennedy promised to conquer the moon before the end of the decade, he was setting a serious challenge. A huge amount of work lay ahead. The enormous cost of such a programme might also be questioned, as sending a man into space – never mind landing on the moon – also entailed bringing him back again. A simple probe could accomplish almost as much, at a hundredth of the cost. But therein lay the essence of the argument: a probe is not a human being and the conquest of the moon would provide new pioneers in an America seeking a new frontier.

It was a time when every flight was featured on magazine covers and *Paris-Match* was full of fascinating double-spread photographs of the events. But after the euphoria, the United States woke up. The dream was over. The Russians – perhaps more pragmatically, but also because they had been overtaken – abandoned the race to the moon to concentrate on manned orbiting space stations. Indeed, since the 1970s manned space flight has been devoted to this type of programme, albeit with the novelty of international co-operation on flights and technology. As for the Shuttle, the delivery-van of space exploration, it has not proved an ideal vehicle. Expensive and in some ways inadequate (two have been lost), it will soon be taken out of service.

The fresh challenge for the United States is the conquest of Mars, scheduled for around 2030, preceded by another trip to the moon. This programme may well provide new inspiration for the world's leading power, yet it will surely be undertaken in partnership with other countries. A new adventure for man in space is quietly getting under way.

ASTRONAUT
JOHN GLENN

CHAPTER 1

The first rocket launched from Cape Canaveral in Florida on 24 July 1950, Bumper was merely an improved German V-2. Other Bumpers had already been fired in the New Mexico desert. These rockets served as test-beds but did carry a variety of equipment. Linked to the second stage of a WAC Corporal (a rocket that had been under development since 1945), Bumper could carry a payload of 700lb into the upper atmosphere.

Two important planes in the American military and NACA programmes: the Bell X-1, the first plane to break the sound barrier in October 1947, and the Douglas Skystreak. The first was financed by the then new Air Force, while the second was built using Navy funding. While the public will remember the supersonic Bell, the D-558 played an important role in the exploration of transonic speeds. On the other hand, the X-1, with its rocket engine, cleared the way towards higher speeds.

Test programmes

The modern rocket, as a launcher of spacecraft, has its roots in a weapon of war developed by Nazi Germany. Certainly the earlier work of pioneers, such as Goddard for the Americans and Tsiolkovski for the Soviets, cannot be dismissed, but it was principally the A-4 rocket – better known as the V-2 ('retaliation weapon number 2') – that, following the end of World War 2, gave impetus to the conquest of space. Although this ballistic missile had no bearing on the outcome of the war, it was nevertheless the 'womb' that gave birth to the whole American and Soviet space programme. Indeed, the Allies seized a large number of these weapons and conscripted most of the engineers who had worked on their production. Wernher von Braun – who took the United States to the moon – was the father of the V-2. Various other specialists, such as Helmut Gröttrup, were captured by the Soviets and 'invited' to work for them, although it is still not known if they played a central role in the Soviet space programme. The improved American and Russian versions of the German A-4 rocket were chiefly intended to be military vehicles and soon came to be known as 'intercontinental ballistic missiles'.

The United States had possessed atomic weapons since 1945, and the Soviet Union also became a nuclear power in 1949. Three years later, an American thermonuclear bomb was exploded at Bikini Atoll and a few months later the USSR carried out its own first test. Possessing such ultimate weapons was one thing, but having the means to deliver them was another. Until the start of the 1950s, the American Air Force's SAC (Strategic Air Command) and the Soviet DA (*Dal'naya Aviatsiya*, or long-range flight) had only bombers, whose limited range left them unable to reach many big cities. It is true that the USAF was about to get the B-47 (followed by the formidable B-52, which would allow it to reach any point on the globe), but in order to achieve target saturation it was clear that the rocket would be the most effective delivery vehicle for an atomic bomb, its relative inaccuracy being offset by its destructive power. Very soon the armed forces of both camps were seeking to equip themselves with reliable launchers. The launch of Sputnik 1 would intensify the quest.

NASA: a hurried birth

On the diplomatic front, the end of the 1940s saw tense relations between the major powers. From 1947, President Truman's doctrine of 'containment' was clear: the expansion of Communism must be halted. For the Soviets, Jdanov's response was that henceforth the world was bi-polar, divided between the imperialists and the anti-imperialists. After the Berlin crisis in 1949 came the Korean War, which did nothing to relieve tensions. The creation of NATO in 1950 and the Warsaw Pact five years later perfectly illustrated the military rivalry between the United States and the USSR.

The Cold War also moved into space, with NASA in the West and its Russian equivalent, the OKB-1 research department directed by the talented Korolev, in the East. Indeed, NASA was specifically set up to counter the Soviet Union in space, so that 'the American people do not go to bed under the light of a Communist moon,' according to Vice-President Johnson. However, before NASA (National Aeronautics and Space Administration) there was NACA (National Advisory Committee for Aeronautics), a specialised aeronautical

UNITED STATES
109D

The pilot and future astronaut Neil Armstrong in the cockpit of his X-15 in 1961. The X-15 programme was very important to NASA. The North American-made plane was intensively used for nearly ten years in space research, as it was able to fly at mach 6 and reach altitudes of more than 50 miles. Attached underneath the wing of a Boeing B-52, it was launched at a high altitude (around 35,000ft) and then fell more than 5,000ft before firing its engines. In July 1962 an X-15 reached an altitude of 59.6 miles, its pilot Bob White thereby becoming the first pilot/astronaut in history.

research laboratory whose work on aerodynamic phenomena was a decisive factor in the history of aviation, especially during World War 2. But NACA, under a limited budget, worked for the American Army (under whose aegis the Air Force still was) as well as for the Navy, the latter having a powerful aviation force of its own.

When the Air Force became independent in 1947 the country found itself with a plethora of aeronautical research departments: those of the Army, the Air Force and the Navy, as well as NACA itself. Under pressure, the first and second of these were soon merged, and questions began to be asked about what further use there could be for NACA, which, though it had done successful work on military aircraft, seemed to be in search of a role after 1945. Certainly, record-breaking supersonic flights had had popular appeal – in October 1947, Chuck Yeager broke the sound barrier in the rocket-engined Bell X-1, and transonic flight testing was made possible by the turbojet Douglas D-558 Skystreak – but NACA was closely dependent on the military (the X-1 was funded by the Air Force and the Douglas D-558 by the Navy). The advent of rockets would give NACA a renewed front-line role thanks to its forceful boss, Hugh Dryden, who cleverly convinced the Air Force and the Navy to put their hands in their pockets to finance, among other things, the prestigious, hypersonic X-15 programme designed to explore the upper layers of the atmosphere, a project of no obvious use to the military! NACA also benefited from the support of President Eisenhower, who was anxious to develop intercontinental missiles to guarantee the nation's security.

Several other military and non-military bodies also got involved in rocket development, including the Pasadena, California-based Jet Propulsion Laboratory (JPL) that had undertaken a lot of research with the V-2s brought to the US after World War 2, and was receiving a budget from the Army to develop launch vehicles that resulted in the WAC Corporal and the Redstone rockets. The Army itself also had a missile research department, the Ballistic Missile Agency, at Huntsville, where Wernher von Braun worked. Von Braun, father of the German V-2, had been working on launchers for nuclear weapons, but the dream he truly harboured was to send a man into space. Well before the Apollo programme, von Braun had been laying foundations for the Saturn rocket. As for the Navy, it had the Viking rocket (the only launcher that was not conceived as a ballistic missile) that had been developed by von Braun's rival, Milton Rosen. Finally, the Air Force was carrying out early tests on its Atlas missile, which was destined to become the first operational, truly intercontinental missile.

The launch of Sputnik 1 really got matters moving, and resulted in the creation of NASA in 1958. Gifted with a huge budget and bringing together JPL, the military research groups and incorporating the Huntsville agency, NASA was ready to conquer the stars.

Without any doubt, Wernher von Braun is the father of space exploration, thanks to the remarkable Saturn launcher that he designed. Unlike Korolev, von Braun played only a minor role in the development of spacecraft, devoting himself almost entirely to rockets. Nevertheless, his influence on the American space programme was fundamental, as he also came up with the idea of space stations.

The Soviet research departments

The Soviets, thanks to their planned, nationalised economy, had developed a rigorous body of research departments who were charged with developing various experimental prototypes, presenting them to the members of the Supreme Soviet and the military, and – once they received backing – producing them. These OKB (*Opytnoe Konstrucktorskoe Byuro*) departments were led by a chief engineer whose name identified the bureau in question. Thus, OKB-1 was run by the engineer Sergei Korolev. It had been set up near Kaliningrad in 1950, emerging out of department 3 of bureau NII-88 (scientific research institute 88) created in 1946, which had been charged with developing and producing a ballistic missile from the seized German V-2 rockets. This barely improved V-2 would give birth to the R-1 rocket. The head of NII-88, a certain Tritko, put Korolev in charge of OKB-1 and in 1956 it finally became independent.

Although OKB-1 had overall supervision of the programme, it worked in close connection with other bureaux, such as Glushko's OKB-456 (engines) and Pilugin's NII-885 (avionics). In fact OKB-1 was not even the only department in its field: Chelomei's OKB-52 and Yangel's OKB-586 would become rivals to Korolev. Each of them had patronage from within the Politburo, which would sometimes cause friction, but all three played their part in the Soviet space effort.

After producing the R-7 launcher, the first intercontinental missile (which had put

The launch of an Atlas-A missile in 1958. These were military missiles that were initially used in space exploration. The Atlas was the first operational American ICBM and had a range of 6,800 miles. Built by Convair, the early Atlas-A missiles had only two booster engines. The modified D version was used to put the Mercury craft into orbit. Atlas was also attached to other stages (Agena or Centaur), and later gave birth to the Atlas I, II, III and IV, which are still in service.

Born in the Ukraine in 1907, chief engineer Sergei Korolev worked in the OKB-1 research department, where he developed the R-7 rocket and the Vostok spacecraft and its successor, Voskhod. He then moved on to the Soyuz project, but died in 1966 during surgery.

The first artificial satellite was Soviet. Sputnik 1 was successfully launched by an R-7 rocket on 4 October 1957. Hurriedly designed by Korolev as a last-minute replacement for a larger satellite, Sputnik was a 184lb aluminium sphere with a transmitter inside that sent out its famous 'beep-beep' signal over a period of 21 days. It remained in orbit until the beginning of January 1958.
(Photo: DR)

Sputnik 1 into orbit in 1957), it was Korolev's bureau that received all the attention. As one can imagine, all this work was top secret and the Soviets imposed an almost complete silence on their programmes. They revealed a scanty outline of Vostok, but only very late in the day after the programme was completed in 1965. Generally speaking the Soviet blackout on their space flight activities was total until quite recently. For example, the existence of the R-7, a rocket that was produced in quantity, wasn't revealed until 1967, ten years after its first flight (although Western observers had managed to get a fairly clear picture of it). The limited information and pictures available contributed to the growth of myths and misinformation – indeed, according to some Gagarin was not the first man in space. Because of the chronic lack of information and reliable sources, the Western press came up with some Soviet spacecraft that were pure fantasy. The Russians were equally keen to cover up their failures, as these might have been seen as incompatible with the Communist notion of progress. Cosmonauts who died during training have only recently received any mention.

On 6 December 1957, Vanguard TV3 explodes a few seconds after launch. The Vanguard series was worse than a disaster, it was a defeat, as out of 12 launches only three were successful, putting into orbit satellites of between just 3lb and 50lb!

NASA's first seven astronauts pose together for posterity. Within a few months, without having set a foot in space, they would become the darlings of the burgeoning American TV media and news magazines. The photograph was taken in July 1962 and shows, back row (left to right) Shepard, Grissom, Cooper; front row (left to right) Schirra, Slayton, Glenn and Carpenter. All of them were Air Force, Navy or Marine Corps pilots.

Astronauts and cosmonauts

Even before the start of the space programmes, the world had been gripped by a disturbing 'space-mania'. In the United States in the early 1950s there was a spreading belief in the existence of little green men. UFOs were so popular that the US Air Force, with great seriousness, set up a commission to look into them (the Blue Book project). Interstellar space was in vogue and works of science fiction – in book, comic strip or movie form – had numerous imitators. This phenomenon would later play an important part in space exploration, as American public opinion thus fed was much more willing to support its government's space programme. Furthermore, volunteers were needed! After the War, it was considered vital to concentrate on technology if success was to be guaranteed, as human beings were by nature too unreliable. Indeed, the (American) astronauts and (Russian) cosmonauts were viewed more as human guinea pigs than pilots fulfilling a mission: Gagarin did not have manual control of his flight, and the seven Mercury-programme astronauts had difficulty getting their point of view heard.

Far right is the R-7 intercontinental missile that was used as a launcher for the Sputnik series (the picture shows Sputnik 2 upright on its launch pad). Sputnik 2 weighed only a little over 1,100lb and the launcher was quite powerful enough to put it into orbit. These early orbital rockets were designated either R-7 or 8K71. For Vostok 1 (left) the R-7 had to be modified with a third stage and thus became 8K72, also known as the 'Vostok rocket'.

Technical details

Name: Vostok 8K72K

Type: orbital launcher

Country: Soviet Union

Manufacturer: Korolev

First launch: 22 December 1960

Last launch: 10 July 1964

Power on lift-off: 3,894.25kN

Engines:
- **First stage (0):** 4 RD-107 rocket engines running on a liquid oxygen/kerosene mixture giving 97 tons of thrust for 118 seconds
- **First stage (1):** 1 RD-108 rocket engine running on a liquid oxygen/kerosene mixture giving 91 tons of thrust for 248 seconds
- **Second stage (2):** 1 RD-0105 rocket engine running on a liquid oxygen/kerosene mixture giving 5.47 tons of thrust for 365 seconds

Height: 101.18ft

Payload: 10,425lb

Total weight: 276 tons

The Vostok launcher

Vostok was sent into space courtesy of the well-known R-7 rocket, to which a third stage had been added. The R-7 was born of the Soviets' need for a suitable carrier for nuclear weapons. Unlike their American counterparts, Soviet bombers were neither sufficiently numerous nor effective for strategic bombing, and it was as a result of this that the Soviets committed themselves to building rockets. The first true missile that could be armed with a nuclear warhead was the R-5 rocket, capable of delivering a 1.4-ton warhead a distance of 750 miles, a range sufficient to hit Western Europe, but not North America. There was thus a need to develop a more powerful missile, which is how the R-7 Semiorka rocket came into existence. The specifications were revised several times, as the weight of the nuclear warhead rose from 3 tons to 5.5, and the development of the engines gave Glushko's research department quite a lot of trouble. It was the very first intercontinental missile deployed by the Soviet Union, from 1959 to 1968, and was normally armed with a 3-megaton nuclear warhead.

Designated SS-6 Spanwood by NATO, the R-7 was 111ft long and weighed 270 tons. It consisted of two stages and was propelled on lift-off by four RD 107 rocket engines running on a liquid oxygen and kerosene mixture. It had a range of 5,000 miles. The first tests began in 1957, from the Baikonur base.

After two failures, the R-7 – also known by its Soviet builder as the 8K71 – managed to travel 3,700 miles on 21 May 1957. A modified version (the 8K71PS) launched Sputniks 1 and 2 in the autumn of 1957. However, from the military point of view, despite the arrival of the R-7A (8K74) with an increased range (7,500 miles), the rocket was a partial failure. With liquid oxygen propulsion, it required about 20 hours to prepare for a launch and the enormous size of the launch site made it easy to detect (the rocket was assembled horizontally, brought to its launching pad, then raised into the vertical position before filling it up with liquid oxygen and kerosene). For a rocket that was supposed to guarantee the nation's security 24 hours out of 24, this was hardly ideal!

Although the R-7 was quickly withdrawn from the Soviet arsenal, its derivatives remained in service as launch vehicles for many years. Vostok 8K72K was sent into space by an R-7 rocket fitted with a third stage specifically designed to launch the spacecraft. Unlike the Americans, who tested several types of launcher during the Mercury programme, the Soviets used only this rocket throughout the Vostok programme. It is worth emphasising that out of 13 launches – none of them manned – only two were failures, and with this 85 per cent success rate Vostok 8K72K was the most reliable launcher of its time. In an improved version, it remains in service today.

The six Vostok-programme Soviet cosmonauts who went into space were, top (left to right) Gagarin, Titov, Nikolayev; and bottom (left to right) Popovich, Bykovsky, Tereshkova. Nelyubov, missing from the photographs, was part of the programme but did not fly. He committed suicide in 1966 after several months of depression. Valentina Tereshkova, the first woman in space, was chosen by Krushchev himself.

Finding volunteers

The Soviets were naturally encouraged by the success of Sputnik and the speed and efficiency with which the programme was managed (taking just one year). Following this period, Korolev and his OKB-1 department had to face competition from other bureaux that were developing a number of projects, but he eventually received the order to design a craft capable of fulfilling a range of missions, civil, military and manned. From around a hundred Red Army pilots considered for the Vostok mission, only eight were finally chosen, and under conditions of the greatest secrecy. Selection and training took place quite rapidly (bearing in mind what little work the first man in space actually did, they did not really require lengthy training). These future cosmonauts had to be at least 30 years old, 5ft 7in tall, weigh no more than 154lb and, of course, be qualified in flying jet planes. During the summer of 1959 various specialists went round the military air bases seeking candidates. The men chosen remained part of the military, but henceforth they made their career within the select group of cosmonauts. Selection began on 6 January 1961 with the formation of training group 1.

Korolev – who supervised the programme – wanted to have 20 men available, but only 12 of them actually 'survived'. Most of the eight who were lost were either injured in training or rejected for disciplinary reasons, but one trainee was killed. Eventually only six pilots passed all of the initial tests. At the head of this new unit was General Kamanin (who would occupy the post until 1971). Gagarin was held in high regard, just behind Titov, whose first name (Gherman) had the disadvantage of sounding a little too much like 'German'. Titov was keen on women, cars and alcohol, so it was not surprising that the Soviet hierarchy chose to relegate him to second place. Gagarin, on the other hand, was a perfect son of the Soviet people: born in 1934, he joined the air force in 1955, became a good pilot and was married. He was finally selected to be a cosmonaut in February 1960, along

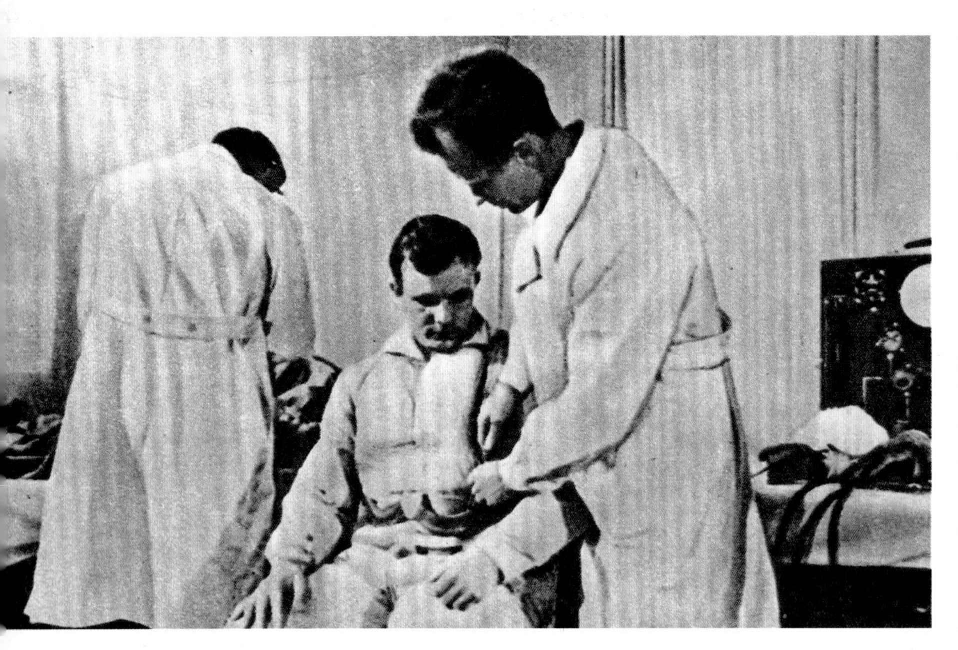

The men in the first body of Soviet cosmonauts underwent a whole battery of tests during their preparation for orbital flight. However, bearing in mind the little they had to do and the short duration of the flights, they did not have to spend long months preparing, as was the case with later missions.

with five other pilots from the Soviet Air Force, and his considerable talents led to him being chosen to be the first man in space. This, however, proved to be something of a poisoned chalice, as, having become a national (and international) hero of the Soviet Union, he was not permitted to fly again until 1966, when he was selected for the Soviet moon programme, which was ultimately abandoned. Sadly, he was killed during a training flight in a MiG-15 on 27 March 1968.

With Gagarin and Titov, the four other cosmonauts selected in 1960 were Nelyubov, who never flew and sank into depression and alcoholism before committing suicide in 1966; Bykovsky, who flew on Vostok 5, Soyuz 22 and docked with Salyut 6; Popovich, who was on board Vostok 4, then 12 years later on Soyuz 14; and finally Nikolayev, who flew on Vostok 3 and Soyuz 9. A four-strong female cosmonaut group was also established in 1962, of whom Valentina Tereshkova would become the best known. She was not from the military, but her past as a worker in the textile industry, as well as the heroism of her father – killed during the Great Patriotic War – and the fact that she was a skilled parachutist, were determining factors in the eyes of her superiors.

In the United States, the Mercury project was making tabloid headlines even before its success. Though confident of its ability, NASA was aware that the venture carried risks and that the conquest of space could be peppered with disasters. Nonetheless, the Americans had no shortage of volunteers seeking to go into space. Recruitment was initially somewhat haphazard, but President Eisenhower intervened to ensure that selection was made from among military test pilots (the Air Force, the Navy and the Marine Corps). The selection procedure was entrusted to a committee directed by Charles Donlan, a little-known engineer and one of the pillars of NASA. He was partnered with Warren North, a test pilot, and various doctors, surgeons, psychiatrists and psychologists. The members of the committee determined that the experiences a man would undergo in space were not very different from those of a military pilot – hence the decision to draw the nation's astronauts from its pool of fighter pilots. The candidates needed to be less than 40 years old, no more than 5ft 10in tall, but with at least ten years of experience and a minimum of 1,500 hours of jet-plane flight. Note that the Soviets picked pilots who were both younger and shorter! In fact, NASA did not intend that its astronauts should make a career in space flight (unlike the Russians), and the selection committee preferred more mature, less easily influenced men. As for the

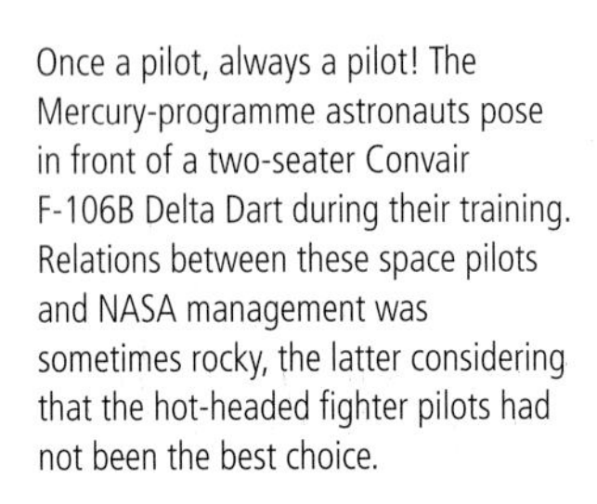

Once a pilot, always a pilot! The Mercury-programme astronauts pose in front of a two-seater Convair F-106B Delta Dart during their training. Relations between these space pilots and NASA management was sometimes rocky, the latter considering that the hot-headed fighter pilots had not been the best choice.

A Vostok craft has been installed under the launcher's nose cone. It will later be mounted on the main body of the rocket. The small open hatch gives access to Vostok. The rocket's maker was Korolev's OKB-1, which had overall responsibility, but a total of 36 organisations contributed to its development. Among them were Isayev's OKB-2, entrusted with the retro-rocket system; Alexeyev's factory 918, which worked on the spacesuit and its accessories; and TSKB-598, which provided the Vzor optic visor.

maximum height, that was readily explained by the 'XL' size of the sons of Uncle Sam!

Though 110 men were originally chosen, only seven were left after a rigorous series of tests. These 'original seven' – a name chosen with the original Seven Wonders of the World in mind – were a focus of interest in the American press even before they made a single flight. Officially presented to the public on 9 April 1959, Carpenter, Cooper, Glenn, Grissom, Schirra, Shepard and Slayton undertook intensive training at the Langley base before being transferred to Houston. Of the seven, only Slayton did not fly, following a cardiac exam – he had to wait 16 years before taking part in the last Apollo /Soyuz mission – but he directed the astronaut group in all the Gemini and Apollo missions. Walter Schirra was the only one to participate in all three programmes, Mercury, Gemini and Apollo. As for Alan Shepard, he was lucky enough to be one of those who set foot on the moon in the Apollo 14 mission in 1971. John Glenn came to be the leader of the group, not because he was the best, possibly because he was the eldest, but above all because of his eloquence in front of the media, which the others lacked. He thus became the spokesman of NASA's Group 1, perfectly balancing the astronauts' interests with those of the Mercury programme as a whole.

The search for a launcher

Vostok ('East') was, of course, the first programme to succeed in sending a man into Earth orbit. The Vostok craft grew out of the Zenit reconnaissance satellite project, which was being developed simultaneously. As for Vostok's launcher, this was simply a development of the R-7 intercontinental missile, as indeed were all the early launch vehicles, whether Soviet or American. Work got under way in the spring of 1957, through research undertaken by the engineer Tikhonravov – under Korolev's leadership – to develop the 'section 9 project'.

In less than a year, the specifications had been worked out. Even though we still look upon the Soviet craft as being rather basic, the first Vostok was cleverly thought out and had an exemplary modularity and versatility, both being hallmarks of Soviet spacecraft. In broad outline, the Vostok launcher was to have a maximum weight of 5.4 tons, be capable of withstanding temperatures of up to 3,500°C and forces of 9G. The rocket itself had to propel a weight of 4.9 tons into low Earth orbit, using a third stage. The design of the spacecraft was revised many times before settling on a simple sphere, which presented the same aerodynamic properties at any speed and under any angle of re-entry into the atmosphere. In short, it was simple and effective.

In June 1958 Korolev himself took over the project, which left the drawing board to go into construction on 15 September 1958. However, the Red Army tried to interfere in its development, causing delays, and it was not finally ratified until May 1959, after some lively exchanges. Vostok was now capable not only of carrying a cosmonaut, but also of being used, in its Zenit guise, as a military spy satellite.

The Vostok launcher, alias 8K72, in the form in which it was unveiled at the Paris Air Show in 1967. Before this date practically nothing was known about this rocket, which had sent Sputnik into orbit ten years previously! It is seen here while being assembled in the open, some years before. The four boosters are attached around the main body, which has an identical engine. Each of these engines has four main nozzles and two other Vernier engines for course correction.

Period photographs of Vostok and Voskhod are all of mediocre quality and are of little interest. The Soviets did not release any pictures of the Vostok craft, so that the Western press, lacking anything else, had to put up with official, censored pictures portraying 'modern Soviet man' or some such. For several years periodicals would resort to drawings – often highly speculative – to cover the story of the pioneers of space. On the left of the photograph is Gagarin heading for the launch pad on a bus. In the background is Titov, Gagarin's backup.

A Russian in space

Fortified by Sputnik's success, the Soviets were in a position to move on to manned orbital flight without too many problems: remember that Sputnik 2 – launched only a month after the legendary Sputnik 1 – had already carried a living creature into space (the famous Laïka, a female dog that died of hypothermia after a few hours, while its craft continued in orbit for another 162 days). Though Sputniks 3 and 4 were just 'orbiting laboratories', Sputnik 4 – launched in May 1960 – was nevertheless significant in being the first Vostok-series craft to be tested in space. (At this point the 'Vostok' designation had not yet been applied, hence the mission being christened Sputnik 4.)

Sputnik 5 – launched on 19 August 1960 – carried two more dogs, called Belka and Strelka. Both of them survived the trip, which was the first Vostok flight to carry living creatures. In just a few months the Russians had succeeded in putting steadily heavier payloads into orbit: Sputnik 1 had weighed little more than 180lb, whereas Sputnik 6 – launched on 1 December 1960 – had reached 4.5 tons, indicating what very rapid progress they had made. When, on the 12 April 1961, Vostok 1 took off with Yuri Gagarin on board, making one complete orbit, NASA could reply only with a simple sub-orbital flight (by Shepard) on 5 May. Undeniably, in this first leg of the space race the Soviets had established a clear lead.

The intelligent spacecraft

Vostok actually comprised two distinct parts: the single-seat command module and the instrument module. Of spherical shape (to ensure the best volume-to-surface ratio), with a diameter of 7?ft, the module weighed about 2.5 tons. Nearly 1.6 tons of this was accounted for by the refractory material that formed a thermal shield for re-entry into the atmosphere. As the whole surface of the sphere was covered in protective material, Vostok could re-enter at any angle, which was not the case with the American Mercury spacecraft. On the other hand, Vostok, with its billiard-ball shape, was restricted to a ballistic re-entry, rather than a controlled descent. The centre of gravity was situated towards the front, insofar as a sphere can be said to have a 'front and a rear'. During its earthward descent, the spacecraft's weight steadily increased under the Earth's gravity and it naturally turned so that the centre of gravity was at the bottom. As this was not controllable and might act in any direction, its speed on leaving orbit was considerable. In fact, the cosmonaut was subject to a force of up to 8G and the accuracy of his return trajectory was low.

The Vostok vessel as it looked before re-entry. The spherical part housed the cosmonaut. The whole of the rear part, here seen resting on a trolley, is the service module. Vostok separated before re-entry into the atmosphere and was destroyed.

The second part of the spacecraft was the instrument module. Attached to the command module by metal strips, the biconical module, with a maximum diameter of 7.9ft, included a collection of round bottles containing oxygen, water reserves and nitrogen. There were also batteries for powering the radio and other electrical systems. The rear section provided the craft's propulsion via a simple retro-rocket whose sole job was to brake Vostok during its re-entry. The engine could be fired only once. If ignition did not occur, the cosmonaut had to wait ten days or so for the spacecraft to gradually come out of orbit and return to Earth by itself, a manoeuvre possible only because of Vostok's low orbit. Gas-jet rockets permitted control of the craft's roll. Joined, the two modules were 24.1ft long and weighed 4.65 tons (on Vostok 1).

The cosmonaut was strapped to a large, 660lb ejector seat which allowed him to escape from the module if there were any problems at launch, but was principally for use on re-entry, as it was not intended that the cosmonaut should remain on board for what would have been a very rough landing. At around 23,000ft the pilot would eject and the speed of descent was stabilised by a small parachute. At about 13,000ft he jettisoned the seat and returned to Earth by deploying his own parachute. In the event of an emergency landing, the cosmonaut's suit allowed him to float in water and to withstand low temperatures. In addition the ejector seat was provided with a beacon, an inflatable dinghy, rations and a survival kit. When in orbit, the cosmonaut had a porthole with a Vzor sight, which, in the daytime, allowed him to establish his position over the Earth's surface, or, if need be, to attempt a manual landing, this normally being radio-controlled from the ground. He could also make adjustments to his position in one plane, using gas jets, but the navigation systems were simple in the extreme and it was not possible for him to change his orbit. There was no gyroscopic stabilisation and only the solar, ion and infra-red sensors – lined up relative to the Earth's horizon – allowed the cosmonaut to adjust the attitude of the spacecraft. Nevertheless, no less than 1,750lb of testing and control equipment could be carried aboard Vostok 1.

PARIS
MATCH
Nº 628/22 AVRIL 1961/0.80 NF

UN TEMOIGNAGE HISTORIQUE QUE VOUS VOUDREZ GARDER
LE PREMIER HOMME
autour de la terre
nos photos exclusives de Moscou
et le reportage de Cartier retour de Russie

The Soviet Union was an atheist nation but the cult of personality approached deification. Gagarin remained a legendary figure, as the quantity of postcards, stamps and other commemorative souvenirs that swamped the USSR demonstrates. Throughout Europe and the rest of the world his exploit made the front page, despite the absence of photographs of it!

Vostok 1 returns to Earth…without its occupant! Contrary to what the Russians claimed for a long time, Gagarin did not come down inside his craft, but descended by parachute, having previously ejected. Vostok had been designed for a purely ballistic re-entry and its impact with the ground was judged too violent to be endured by man. For the first human spaceflight to be validated, the cosmonaut had to return in his craft. The Soviets only admitted the truth much later.

Gagarin's flight

Wednesday 12 April dawned with beautiful weather over Kazakhstan, and close to Tiouratam (the future Baikonur), a three-stage R-7 rocket stood pointing towards the sky. Inside the Vostok spacecraft carried under the rocket's cap, Lieutenant Yuri Gagarin was preparing himself. A few hours earlier, he had arrived at the base with Titov, his stand-in, who would be ready to replace him at short notice if he was sick. But Gagarin was in top form. When Korolev told him that the countdown was about to begin, the cosmonaut simply replied: 'Understood. I feel perfectly fine and in excellent spirits.' At 06:07 the four engines of the R-7 Vostok 8K72K rocket burst into life with a deafening roar. Fighting against the pull of the Earth's gravity, it began to accelerate. *'Poyekhali!'* ('We're off!') cried Gagarin. At 06:09 the four lateral boosters of the first stage cut out and the second-stage engine continued its trajectory. Then the cap was ejected and the Soviet cosmonaut could at last see outside: 'I can see the Earth and visibility is very good,' he reported laconically. At 06:12 the main stage separated and the final stage took over. At 06:14 Gagarin called out: 'Everything's going well, let's keep it up!' Then he realised that communications with the ground were deteriorating as he moved out of range of the main transmitter. By 06:17 the final stage had cut out and Vostok was in a stable orbit. In 108 minutes it would complete a full circuit of the Earth at a speed close to 17,500mph.

Though well prepared the mission still carried numerous risks, as it was obviously unknown how the human body would cope in space. Furthermore the engineers had locked the controls, so that Gagarin could not immediately take manual charge of the flight. The codes to unlock them – handwritten – were in an envelope that the cosmonaut could open, if he needed to… After an hour the spacecraft automatically put itself in position for re-entry. This was the most delicate moment of the return flight, as the retro-rockets that manoeuvred the craft were crucial to putting it into the correct re-entry orbit. If the ignition did not work, there was little chance of Gagarin coming back. The orbit was quite low (with a perigee of only 109 miles), and had the retro-rocket failed Vostok would have undergone a natural, aerodynamic

Far left: A view of the Baikonur launch centre during the Vostok era. The wooden chairs and tables are standard, as are the uniforms. The centre is situated in Kazakhstan, not far from the Aral Sea (or what remains of it). Built in 1955, the cosmodrome is still in service, but as it is now on territory that no longer belongs to Russia the Russian Confederation rents it from Kazakhstan.

Left: Popovich poses on the Vostok 4 rocket's launch pad. Pavel Romanovich Popovich was aged 32 when he was accepted into the first cosmonaut group. From the published photographs, he always seems to be smiling and in good spirits. On the Vostok 4 mission he flew with Nikolayev, who had gone up the previous day in Vostok 3. The two craft came to within three miles of each other.

slowing down after about ten days (as well as succumbing to the Earth's gravitational pull) and would have re-entered the lower layers of the atmosphere, but with no way of knowing where it might come down.

The descent from orbit began over Angola, about 5,000 miles from the expected landing point. The rockets fired correctly for around 40 seconds, but the instrument module did not separate from the command module as it should have done after a ten-second burn, and this started to destabilise the craft. The problem was caused by one of the strips holding the two modules together, but it eventually burnt up and the command module was able to continue its re-entry alone. After withstanding almost 8G at an altitude of 23,000ft, Gagarin ejected from the module as intended and regained land about ten minutes later. As for Vostok 1, its parachutes did not deploy until 8,000ft and it hit the ground quite violently, bouncing at least once.

When Gagarin himself reached terra firma he was greeted by a peasant and his daughter, who were much intrigued by this man in an orange suit, dragging a parachute behind him! 'Don't be afraid,' he told them, 'I'm a Soviet, just like you.' Then, for good measure, he added: 'I've come from space and I need to find a phone to call Moscow.' Within a few days, Gagarin had become a hero, a true standard-bearer of the triumphant Soviet regime. But was he really the first man in space? Not quite. For a manned space flight to be recognised according to international convention the cosmonaut had to return in his spacecraft, and, for many years, the Soviets hid from the world the fact that he had ejected... Likewise, Vostok, with its spherical shape, was not revealed to the West until 1965.

The other Vostoks

The Russians did not stop there and, up to June 1963, no fewer than five other flights took place. All were successful. Vostok 2 was launched on 6 August 1961 from Baikonur with Gherman Titov on board. Only 26 years old (he has long been the youngest cosmonaut in history), Titov was also the first to spend a whole day in space, achieving 17 ½ orbits. During this flight the pilot was able to take manual control of the module for a few minutes. The return to Earth passed without mishap, but, as with Vostok 1, the two modules did not separate properly. Titov also ejected, but the command module was destroyed on hitting the ground.

An R-7 rocket being assembled. With the R-7, the Soviets had the most modern rocket there was. It was an R-7 that launched the Sputniks, Vostoks and Voskhods, not to mention numerous military satellites.

An envelope dated 3 March 1962 on the occasion of John Glenn's homecoming to the town of his birth, New Concord, Ohio.

Vostoks 3 and 4 took off on 11 and 12 August 1962. For a few hours the two spacecraft were in flight together, three miles apart, and the two cosmonauts (Nikolayev and Popovich) were even able to make radio contact with each other. Vostok 3 remained in orbit for almost four days, as against three for Vostok 4, which returned to Earth earlier than planned (ground control thought that they had received a signal from Popovich to abort the mission). Nikolayev was the first man to take colour photographs of the Earth from space.

Finally, ten months later, Vostoks 5 and 6 (with, respectively, Bykovsky and Tereshkova, the first female cosmonaut, on board) carried out the same mission as Vostoks 3 and 4. Once again, despite a few concerns, the two missions went very well, although Tereshkova did not particularly enjoy her 2 hours 22 minutes in orbit, suffering the whole time from space sickness. It had been originally intended that two women would fly in the two spacecraft, but the cosmonaut Ponomaryova, who was to have taken the controls of Vostok 5, showed an independence of mind that did not go down well with the Soviet hierarchy. Consequently despite being capable and motivated, she never flew in space. The Soviets scored points by sending a woman into space, and at an international level Tereshkova's flight had positive repercussions. Even though it has to be said she was chosen for purely propaganda reasons, she nevertheless became a spokesperson for women across the globe and even received the title of 'Woman of the Century' in the year 2000.

Mercury strikes back

For the Americans, Gagarin's flight was more than a slap in the face. Sputnik had already made an impression on public opinion in 1957, and here was a Communist, in a steel ball, calmly thumbing his nose at the leading Western economic power. Quite apart from the human and technical achievement, Vostok had also called into question Uncle Sam's military might. Seven years after the Korean War and a few months before the Cuban Crisis, it was clear that the Russians had a complete mastery in rockets and missiles. The Vostok missions 1 to 6 were real successes that damaged the young NASA. The Redstone rocket, ready in 1953, had become operational between 1955 and 1958, so it wasn't as if NACA (NASA's predecessor) hadn't had a launcher available.

The Mercury programme (1959–63) was America's mission to send a man into space. Not all the spacecraft on these missions were actually manned, but all used the Mercury capsule. Very cramped (the Mercury astronauts used to say they 'wore' their capsules!), Mercury was designed for flights of very short duration. Its production was entrusted to the McDonnell Company. The spacecraft was quite different from the Russian Vostok in almost every way, its conical shape being the most obvious. Rather than two modules, there was just one, the section housing the retro-rockets remaining attached during re-entry. The whole of the rear face served as a heat shield, so Mercury had to make re-entry with this side forward, otherwise the capsule and its occupant would have burned up. Mercury was turned into the correct position for re-entry by a folding flap, or spoiler, located at the base of the cylindrical nose, which also held the descent parachutes. The ambient air pressure on this flap turned the capsule into the right position. To come out of orbit, three solid-fuel rockets, fired for ten seconds each, were considered adequate, before being jettisoned upon final re-entry.

There was no ejector seat fitted, so in the event of a launch failure the astronaut's salvation depended on the LES (Launch Escape System), an escape tower mounted on a tubular edifice above the spacecraft. A solid-fuel rocket would fire for one second, separating the capsule from the launch vehicle. The tower was later cast off and the

This page: Testing and fitting of Mercury's retro-package system. Note the straps which hold the retro-rockets in place during re-entry into the atmosphere. The heat shield itself was then jettisoned to allow inflation of the airbags that would keep the capsule afloat when it hit the water. Note also how small the craft is, being used only for very short flights.

spacecraft's parachutes deployed. Like Vostok, Mercury was unable to change its orbit, but it could adjust its attitude using small rockets situated on the sides and the nose. The American capsule was also notably lighter (less than 2 tons on lift-off, as opposed to Vostok's 4.6 tons) and therefore cheaper to build. On the other hand, its smaller size allowed little further development over the ensuing years. After the capsule had been developed, it had to undergo numerous tests before it was considered ready for manned flight. While the Russians used dogs as guinea pigs, the Americans set their store by rhesus monkeys and chimpanzees. In all, 20 exploratory flights were undertaken with several launch vehicles and, on 29 November 1961 the chimpanzee Enos, without realising it, completed two orbits before returning safe and well to Earth. In the meantime the Mercury capsule had carried out two sub-orbital flights with Alan Shepard and Virgil Grissom.

Technical details

Name: Redstone MRLV

Type: sub-orbital launcher

Country: United States

Manufacturer: Chrysler Corporation

First launch: 20 August 1953

Last launch: 30 November 1965

Power on lift-off: 347kN

Engine:
1 North-American Rocketdyne NAA75- 100 (A-6) running on a liquid oxygen/liquid alcohol mixture giving 369kN of thrust for 143 seconds

Height: 69.23ft

Payload: 3,080lb

Total weight: 27.3 tons

The American launchers: a mixed arsenal

Unlike the Europeans and the Soviets, the Americans had not really backed the rocket engine and it was not until the end of World War 2 that they began to show an interest. They were in possession of the leading expert in the world at this time in the person of Wernher von Braun, inventor of the sinister German V-2s. Several of these captured weapons were brought back to the United States and some of them were fitted with a second stage. Given the name 'Bumpers', these rockets served as test-beds. Von Braun was 'invited' to the US Redstone arsenal in Alabama, where he set to work on launch vehicles – and launch vehicles only – which would send America into space and on to the moon. Just like the Russians, the Americans were chiefly developing rockets that could deliver nuclear weapons.

The Redstone rocket, a direct development of the German V-2 and mass produced by Chrysler, was a MRBM (Medium Range Ballistic Missile), the equivalent of the Soviet R-5, but with a much shorter range than the R-7 and smaller in size. Furthermore, the complexity involved in its deployment – it needed 20 vehicles to transport the rocket, its payload, and all the equipment necessary for launch – and the length of time it took to prepare it (between 8 and 10 hours), rendered the Redstone militarily obsolete after 1964. Nevertheless, this rocket with its Rocketdyne engine played a vital role, foreshadowing the future Jupiter and Atlas missiles.

The Redstone programme was under the direction of the Air Force and as a launcher it would certainly have been more effective than the Navy-developed Vanguard rocket. But the administration had put its weight behind the Vanguard…and came to regret it. In the end, the Redstone proved to be the only truly operational US rocket by the end of the 1950s and provided the opportunity to test the Mercury capsule in sub-orbital flight, despite its poor payload. More than 100 flights were undertaken between 1953 and 1965, including

A Redstone rocket takes the chimpanzee Ham for a sub-orbital flight to test the Mercury-Redstone concept (mission MR-2, 31 January 1961). The launcher is fired vertically.

Opposite top: An Atlas launcher is unloaded from a transport plane in 1961. The Atlas was a much more powerful rocket than the Redstone. Of military origin, it found its moment of glory with NASA, propelling all the Mercury capsules that went into orbit.

those using the derivatives of the Redstone, the Jupiter A and C, as well as the Juno.

If we exclude the Juno launch vehicle, which was not ultimately used to carry Mercury capsules, the second rocket chosen for orbital flights was the Atlas, a prolific launcher, as it spawned a whole family: Atlas B and D, Agena, Centaur and Atlas I to V. For the period covered by the Mercury programme, the firstborn of the family is the one that interests us here. The Atlas launch vehicle began its career at the end of the 1940s, when the Air Corps was looking to equip itself with a ballistic missile. The Consolidated-Vultee company responded to the challenge and, thanks to its brilliant engineer of Belgian origin, Karel Bossart, was well placed to pursue a new technique, using pressurised compartment construction. The rocket's body was simply made from steel or aluminium sheet a few millimetres thick and formed the fuel tanks. To prevent the whole structure from collapsing under its own weight and that of the fuel, the interior was pressurised. In this way, a very light rocket was obtained, allowing heavy military payloads to be carried.

The contract was eventually awarded to Convair (formerly Consolidated-Vultee) who also experimented with detachable nuclear warheads and gimballed engines, solutions that were later retained on almost all the space launch vehicles. The project, initially designated MX-774 before becoming the Atlas, focused on developing an ICBM, which would become operational from 1963. The programme was accelerated in response to the threat from the Soviet ICBMs, and the Atlas type A achieved a successful flight in June 1958. Powered by two rockets burning a liquid oxygen/kerosene mixture, Atlas type A was replaced by the type B with 1½ stages. The engines did not operate in succession, but at the same time: two boosters of 67 tons thrust and an engine of 25.5 tons thrust were used for lift-off, but the boosters then separated and the main engine increased its thrust to 36.4 tons in the rarefied atmosphere. From the ground, it looked as if there were three identical engines and the rocket had just one stage. In addition, the rocket had side-mounted, 990lb-thrust Vernier engines that were used to make any necessary adjustments to the trajectory.

In the end, it was the Atlas D that was chosen for the first space flights. At the end of the 1950s the Atlas rocket was the largest in the American arsenal, and by 1958 it was clear that it would be the chosen launch vehicle for the Mercury capsule. Eventually, an order for nine launch vehicles was authorised for the whole of the programme (increased to 15 for the 1962 fiscal year).

On 9 September 1959 an Atlas rocket – nicknamed 'Big Joe' – was used to test the Mercury capsule's heat shield. This first launch of a Mercury capsule by an Atlas rocket (programme MA-1) was a failure, the launch vehicle exploding a minute after lift-off. However, the second launch, also unmanned, was a success. The Atlas D on the third flight (with a dummy on board) was less lucky, refusing to adopt the correct trajectory, but the capsule was recovered after being separated from its recalcitrant launcher. The fourth and fifth launches – the latter carrying the monkey Enos – were the first successful orbital flights. At this point, before John Glenn's flight, the launch vehicle had a success rate of 75.55 per cent, a remarkable figure. Atlas later served as the basis for many other launch vehicles, namely Atlas-Agena, Atlas-Centaur and Atlas-5.

Technical details

NASA Designation: Atlas LV-3B/Mercury

Type: orbital launcher

Country: United States

Manufacturer: Convair

First launch: 14 April 1959

Last launch: 7 November 1967

Power on lift-off: 1,587kN

Engines:
– First stage (0):
2 Rocketdyne XLR-89-5 rocket engines running on a liquid oxygen/kerosene mixture giving 67 tons of thrust for 135 seconds

– First stage (1):
1 Rocketdyne XLR- 105-5 rocket engine running on a liquid oxygen/kerosene mixture giving 36 tons of thrust for 303 seconds

Height: 82ft

Payload: 3,000lb

Total weight: 114.2 tons

Two sub-orbital flights

Unlike the Russians, the Americans chose initially to make short 'hops' into space with two men, using the Redstone rocket. The aim was not to make an orbital flight, but to get into space and return safe and sound with the capsule. NASA's engineers were exceptionally cautious. Up to that point, the various launches had been attended by numerous failures and American public opinion had begun to question NASA's competence and the worth of the whole programme.

The first Mercury capsule to be sent up was named *Freedom 7* and in view of its significance it received maximum attention from the engineers. The flight had been planned for the end of 1960, but the modification and change of various parts delayed lift-off until 2 May 1961. Of the seven astronauts in the Mercury project, three had been picked out for this mission (Shepard, Grissom and Glenn), but on the appointed date on the launch pad at Cape Canaveral the name of the man strapped into *Freedom 7* had still not been revealed. Poor weather conditions eventually delayed lift-off until 5 May, when Shepard's name was finally released.

At 05:15 he took his place in the capsule. Usually the preparation for launch would take a little over two hours, but various technical hitches, as well as unexpected cloud cover, extended the time to four and a quarter hours. Forty-five million Americans were glued to their television screens as, after a final countdown, Shepard's pulse leapt from 80 to 126 beats when he heard the words 'Lift off' inside his helmet. It was 09:34, local time, and through *Freedom 7*'s porthole window Shepard saw the umbilical cable and its support boom linking the tower to the rocket fall away. He started the on-board chronometer recording the flight time and let out an 'Aaah!' of relief, followed by, 'Roger. Lift off and the clock has started. I hear you loud and clear. This is *Freedom*, everything is OK.' Shepard found himself surprised at how gentle the lift-off was. Sixteen seconds into the flight, the launcher and its capsule leaned to 45°. Forty-five seconds later he felt powerful vibrations, telling him that he was now into the transonic speed zone. After two minutes and 22 seconds, the Redstone's engine cut out and the escape tower was jettisoned according to plan. Finally, the capsule separated from the launch vehicle.

For a few minutes Shepard took over control of *Freedom 7*. Flying about 180 miles above the surface of the Earth, he could make out part of Florida below him. Five minutes and 15 seconds into the flight he fired his retro-rockets. The capsule stabilised automatically during the descent. Shepard was submitted to a force of around 11.6G before the first parachute opened at 22,000ft. At about 11,000ft the main chute was deployed. With the heat shield jettisoned, its place was taken by an airbag to keep it afloat. After 15 minutes and 22 seconds of flight, *Freedom 7* made splashdown. Once it had settled in the water, it was secured by a helicopter from the aircraft carrier *Lake Champlain* and lifted a metre above the water. Shepard then opened the hatch and was winched up to the helicopter. If one is to go by internationally accepted standards, Shepard was therefore the first man in space, as, unlike Gagarin, he had returned to Earth on board his spacecraft.

The second flight was equally successful, but had a more dramatic end. *Liberty Bell 7* was piloted by Virgil 'Gus' Grissom. The capsule was almost identical to *Freedom 7* but had been given a central, rectangular window in place of the two small, round ones, and a new access and escape hatch fitted with 70 explosive bolts. This hatch could be opened from either inside or outside the capsule. When *Liberty Bell 7* splashed down, Grissom had difficulty in removing his helmet. The helicopter that had come to pick up the capsule waited a few seconds; Grissom then removed the hatch's safety pin, but did not fire the bolts. For some unknown reason the bolts exploded anyway and the hatch was blown into the water. The floating capsule – not yet attached to the helicopter – began to take on water, but Grissom managed to extricate himself while the helicopter hitched itself to the capsule. However, with the water it had taken on board *Liberty Bell 7* proved far too heavy for the helicopter to lift, and the capsule had to be hurriedly jettisoned and quickly sank. Grissom himself, meanwhile, tired and buffeted by the swell, had considerable difficulty in swimming. Even though his watertight suit provided some buoyancy, the air it contained was rapidly escaping when he was finally fished out by a second helicopter. Blamed by some and defended by others, Grissom did not receive all the honours he merited. Yet the whole of America wept when he died tragically in 1967 during a simulation exercise aboard Apollo 1.

Opposite: Alan Shepard made the first American sub-orbital flight. Strapped into his *Freedom 7* capsule, he eventually took off from Complex 5 at Cape Canaveral on the morning of 6 May 1961. Note how little smoke and steam is generated by the Redstone – you can just about see the hot gases coming from the main engine! The Atlas launchers would be much noisier and would require more extensive logistics, as well as a new launch pad.

Shepard experienced the absence of gravity for only a few minutes, as his flight lasted no longer than a quarter of an hour. *Freedom 7* came down in the Atlantic 300 miles from its launch site. The astronaut was picked up by a helicopter from the American aircraft carrier *Lake Champlain*.

Mission MA-4 almost ended badly. At the end of a faultless flight, Grissom had to evacuate his capsule prematurely as it took in water. The astronaut was saved, but the helicopter could not lift *Liberty Bell 7*. Initially the finger of blame was pointed at Grissom, and though a commission of enquiry subsequently showed that he had not failed in his duty he was deeply affected by the episode.

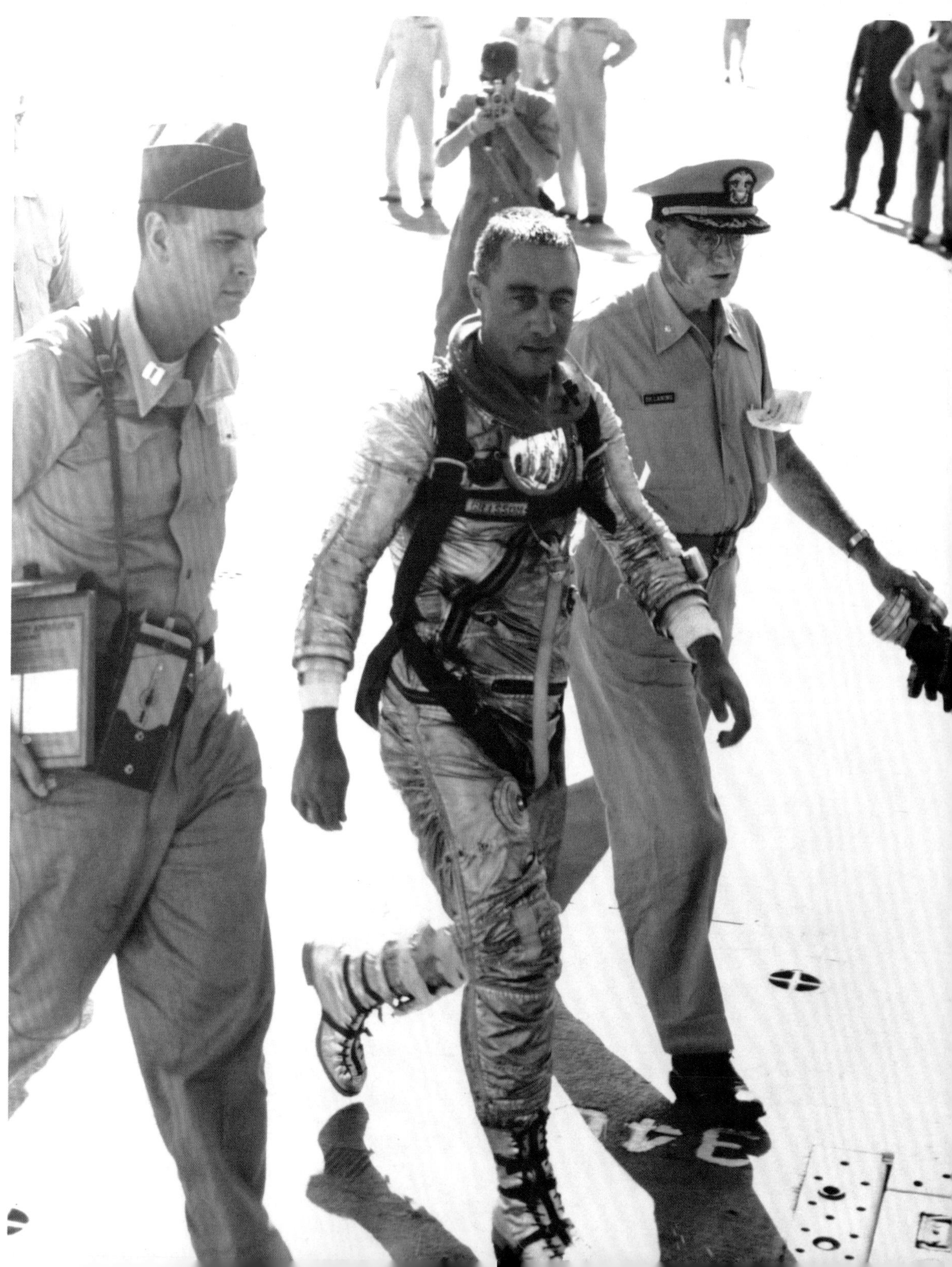

LIFE
INTERNATIONAL
EXCLUSIVE PHOTOS, FIRSTHAND REPORTS
ORBIT!
ASTRONAUT JOHN GLENN
MARCH 12, 1962

John Glenn makes the cover of *Life* magazine on 12 March 1962, a little over a fortnight after his historic flight. Glenn was every inch the American hero – brave (he had served in the Korean War), married and a father.

John Glenn, American hero

Before approving a manned orbital flight, NASA had decided to test its Atlas launch vehicle by sending up the chimpanzee Enos in a Mercury capsule on 29 November 1961. With the success of this mission, nothing stood in the way of a manned orbit. At the end of 1961 a press conference unveiled the names of the future astronauts. By now it was clear that it would not be possible to send an American into space in the same year as Gagarin, and the honour of piloting the first flight fell to John Glenn. Planned for 16 January 1962, the launch was postponed to 23 and then 27 January, before being delayed yet again. It finally took place on 20 February, when John Glenn lifted off in *Friendship 7* at 14:47, local time, from complex 14 at Cape Canaveral. Unlike the other Mercury capsules, launched by Redstone rockets, the Atlas launcher was controlled by a General Electric-Burroughs radio-inertial guidance system, which locked on to a transponder in the launch vehicle. The system allowed the rocket and its capsule to be guided into the desired orbit.

The first stage of the launch went as

Mission MA-6 began very early for John Glenn, when he was woken at 2:20 in the morning of 20 February 1962. After a good breakfast and a medical exam, he put on his G4 suit helped by Joe Schmidt. He had to wait a little longer as the weather forecast was not very good and the Atlas rocket's guidance system had to be reset. At 6:03 he got into *Friendship 7*. As the hatch was about to be closed and bolted, it was discovered that one of the bolts was damaged, incurring a 40-minute delay. A hundred million people were watching their TV screens when, at 9:47 the engines of Atlas number 109-D were fired and roared into life. The guidance system then locked in to the correct course, and after a few minutes Glenn began to feel the effects of weightlessness. He returned to Earth after almost five hours in space. He was picked up dripping with sweat and swung aboard the destroyer *Noa* with his capsule. Now he could enjoy his achievement!

planned: the Atlas's boosters cut out after two minutes and ten seconds, the escape tower was jettisoned 23 seconds later, the sustainer engine stopped after five minutes and 20 seconds, then the capsule separated from its launcher before turning 180° so that the heat shield was facing forward. After a few seconds, *Friendship* had entered the correct orbit and was ready to complete seven orbits at a speed of 25,700ft/sec. Amazed by his view of the Earth, John Glenn passed from day into night above the Indian Ocean. Radio communication was maintained via Australia, with astronaut Gordon Cooper chatting to Glenn for a few minutes: 'That was about the shortest day I've ever run into!' exclaimed Glenn. Indeed, 45 minutes later he was back in the sun as it rose over the Pacific. It was at this moment that he noticed what appeared to be 'fireflies' streaming past his window like thousands of tiny will o' the wisps. No one knew what they were, but it is likely that the 'fireflies' were simply ice crystals illuminated by the rising sun.

After a few minutes Glenn noticed that the capsule's automatic pitch and yaw control was defective, so he changed to manual control. At the start of the second orbit, a ground controller detected an alarm. It appeared that the heat shield and the landing bag were not properly attached, which could have had catastrophic consequences (in fact everything was fine, the alarm having been set off by a defective sensor). Glenn was not told immediately, but he began to wonder why they kept asking him to check that the landing-bag deploy switch was off. Cooped up in his hot space suit and with his attention focussed on piloting *Friendship 7*, he didn't pursue the matter. After completing three orbits he was finally informed of the potential problem with the heat shield. At ground control, the mission director decided to hold on to the retro rockets rather than ditch them as planned. Glenn would also have to maintain manual control during re-entry into the atmosphere.

The intense heat generated by friction caused the retro-rockets to burn up, and Glenn could see incandescent pieces of them flying past his window. *Friendship 7* began to shake from side to side and its occupant remembers having been tossed around 'like a falling leaf'. He decided to open the drogue parachute, but it deployed of its own accord at five miles (instead of the usual four miles). The main parachute deployed fully, the landing-bag opened correctly and *Friendship 7* made splashdown in the Atlantic 40 miles from the intended spot, the destroyer *Noa* locating it and picking it up. Having got himself out of the capsule, one of the first things Glenn said once on board the *Noa* was: 'It was hot in there!'

The first orbital flight by an American was fittingly celebrated in the United States, but the Mercury programme was far from over.

Orbit, did you say orbit?

The strict definition of an orbit is the trajectory followed by a body in space around another, larger body under the influence of gravity. It is drawn from the well-known universal law of gravity known as 'Newton's Law': bodies attract one another in proportion to their mass and, inversely, in proportion to the square of their distance.

Most manned flights have been made in low Earth orbit (from 200 to 875 miles altitude, approximately, as far as the Van Allen Belt), except, of course, for the Apollo programme's lunar missions. For a satellite to gravitate around the Earth, a minimum speed of 4.913 miles per second must be achieved (what is known as the escape velocity); then, to remain in orbit, it has to maintain a speed of around 17,500mph. Unlike a plane flying in the atmosphere, an object in orbit does not move about freely in any direction, but according to circular orbits passing through the poles (polar orbit), the equator (equatorial orbit) or through any other angle between these two trajectories. It cannot follow a random orbit because its orbit is imposed upon it by gravity. An orbit, however, is rarely circular (if only because the Earth itself is not a perfect sphere).

Mostly, an orbit describes an ellipse. An elliptical orbit is defined by two points: one where the orbit passes closest to Earth (the perigee) and one where it is farthest away (the apogee). But whatever the orbit followed, the satellite will end up by falling to Earth as it is gradually slowed by the gravitational pull; which explains why orbiting stations are regularly 'raised' using their engines, unless the time has come for them to be destroyed (as was the case with Mir, for example). To bring back a craft and its crew, it simply has to be braked (using retro-rockets). Once speed has dropped below the 17,500mph threshold, the craft will come out of orbit.

Scott Carpenter (seen here inspecting his *Aurora 7* craft) took off on 24 May 1962 from Cape Canaveral's Complex 14. Carpenter was John Glenn's backup in 1962 and replaced Deke Slayton when he was grounded with cardiac trouble. Mission MA-7 was the first scientific mission. Carpenter would not serve again with NASA after this, but became a major player in a new experiment with Sealab II and Sealab III, a submarine laboratory in which he lived for 30 days.

The programme comes to a speedy conclusion

This first orbital flight was, to be sure, not the last. On 24 May 1962, Scott Carpenter lifted off on board *Aurora 7*. At the last moment Carpenter had replaced Deke Slayton, who had been taken off flights after some minor cardiac problems. The mission had, above all, a scientific focus, and Carpenter had to make observations of the Earth, take photographs of atmospheric phenomena and assess the behaviour of liquids in the weightless environment of space. He was also asked to try and identify the 'fireflies' seen by Glenn. He was also the first astronaut to be given solid food to eat. On all these points, the mission was a complete success. However, after three orbits the capsule completely missed its planned re-entry point and splashed down more than 250 miles from the intended spot because of a problem with the automatic system. In fact Carpenter had to face several technical problems on top of a heavy timetable, which hardly helped his piloting of the spacecraft. Chris Kraft, the flight director, had no qualms about criticising Carpenter. Kraft directed all NASA's flights up to the Apollo missions. Extremely talented, but somewhat authoritarian (no astronaut would

Walter Schirra, the penultimate of the Mercury programme astronauts, successfully piloted his Mercury *Sigma 7* capsule. So as not to repeat the previous mission's mistakes – *Aurora 7* had splashed down a long way beyond its intended spot – Schirra's MA-8 mission would be kept simple.

Gordon 'Gordo' Cooper started his career as a jet pilot before becoming a test pilot at Edwards Air Force Base. He spent over 34 hours in orbit aboard *Faith 7*, more than all the other Mercury missions combined. During the mission he ejected a small flashing sphere that he had to track during the ensuing orbits. When the main hatch was opened on board the carrier USS *Kearsarge*, a smiling Cooper emerged. The Mercury programme was complete.

ever question his decisions), he was the cause of a certain amount of friction, but his sanctions were instant: Carpenter would not fly again, and his clashes with Walter Schirra on the *Apollo 7* mission led to his being dropped from the team.

It was the selfsame Schirra who took off on 3 October 1962 on *Sigma 7* for six orbits and a total flight time of nine hours thirteen minutes. This flight was also one of the furthest from the Earth, with an apogee of 176 miles. Schirra was able to take some excellent photographs, and talk live on radio and TV for the first time. On this occasion the splashdown took place in the Pacific, not far from Midway. The flight of the ninth Mercury capsule, *Faith 7*, piloted by Gordon Cooper, was also the longest. On 15 May 1962 'Gordo' lifted off to complete 22 orbits, the longest of the Mercury programme. By this time many NASA engineers thought that the previous flight had gone so well that it was not worth launching another Mercury capsule and it would have been better to have concentrated on the Gemini programme. But at the same time, *Faith 7* provided experience of missions of 24 hours' duration or more.

Although it had been planned for 14 May, Cooper took off on the 15th at eight o'clock in the morning, local time. He had a busy programme, with 11 experiments to carry out. He took a lot of photographs (including some scenes inside the capsule), ate, slept a bit, and released into space a sphere fitted with xenon flash lights, which he had to spot and track. At the end of his 21st orbit he was confronted with several failures, but he was able to take manual control of *Faith 7* and splashed down in the Pacific after 34 hours 19 minutes and 49 seconds of space flight. The Mercury programme was then halted in favour of Gemini, and the planned Mercury 10 mission with Shepard on board never left the ground. In the meantime, however, the Russians had put on a spurt.

Chronology of the Vostok and Mercury programmes

Mission	Launch date	Comments
Sputnik 4 (Korabl-sputnik 1)	15 May 1960	Unmanned test of Vostok craft.
Mercury-Atlas 1	29 July 1960	Mercury capsule and Atlas launcher test flight.
Sputnik 5 (Korabl-sputnik 2)	19 August 1960	Vostok craft test with various animals on board, including dogs Belka and Strelka.
Mercury-Redstone 1	21 November 1960	Mercury capsule test flight.
Sputnik 6 (Korabl-sputnik 3)	1 December 1960	Vostok craft test with various animals on board, including dogs Pchelka and Mushka.
Mercury-Redstone 1A	19 December 1960	Mercury capsule test flight.
Mercury-Redstone 2	31 January 1961	Mercury craft test with chimpanzee Ham.
Mercury-Atlas 2	21 February 1961	Mercury capsule and Atlas launcher test flight.
Sputnik 9 (Korabl-sputnik 9)	9 March 1961	Vostok craft test with human dummy and dog Chermushka.
Mercury-Redstone BD	24 March 1961	Mercury capsule and modified Redstone launcher test flight.
Sputnik 10 (Korabl-sputnik 10)	25 March 1961	Vostok craft test with human dummy and dog Zvezdochka.
Vostok 1	12 April 1961	First manned space flight, with cosmonaut Yuri Gagarin.
Mercury-Atlas 3	25 April 1961	Mercury capsule and Atlas launcher test flight.
Mercury-Redstone 3	5 May 1961	First American sub-orbital flight.
Mercury-Redstone 4	21 July 1961	Second American sub-orbital flight.
Vostok 2	6 August 1961	Second manned flight, with cosmonaut Gherman Titov.
Mercury-Atlas 4	13 September 1961	Mercury capsule test flight.
Mercury-Atlas 5	29 November 1961	Mercury capsule test flight with chimpanzee Enos.
Mercury-Atlas 6	20 February 1962	First American orbital flight, with astronaut John Glenn.
Mercury-Atlas 7	24 May 1962	Second American orbital flight, with astronaut Scott Carpenter.
Vostok 3	11 August 1962	Third Soviet orbital flight, with cosmonaut Andrian Nikolayev.
Vostok 4	12 August 1962	Fourth Soviet orbital flight, with cosmonaut Pavel Popovich.
Mercury-Atlas 8	3 October 1962	Third American orbital flight, with astronaut Walter 'Wally' Schirra.
Mercury-Atlas 9	15 May 1963	Fourth American orbital flight, with astronaut Walter Gordon Cooper.
Vostok 5	14 June 1963	Fifth Soviet orbital flight, with cosmonaut Valery Bykovsky.
Vostok 6	16 June 1963	Sixth Soviet orbital flight, with cosmonaut Valentina Tereshkova.

Conclusions

Conclusions
Semi-failure, Vostok craft did not enter correct orbit.
Sub-orbital flight, failure, Atlas rocket exploded 58 seconds into flight.
Successful orbital flight. Craft and animals returned to Earth after 24 hours.
Failure, Redstone rocket did not lift off.
—
Sub-orbital flight, complete success.
Successful sub-orbital flight.
Sub-orbital flight, complete success.
Successful orbital flight.
Sub-orbital flight, complete success.
Successful orbital flight.
First orbital human space flight. Flight lasted 1 hour 48 minutes.
Sub-orbital flight. Partial failure. Rocket was destroyed, but capsule ejected, went into its orbit and returned to ground intact.
Astronaut Alan Shepard makes a space hop of just over 15 minutes.
The capsule's first flight into orbit. Mercury carried Virgil 'Gus' Grissom on a 15-minute sub-orbital flight. Capsule was lost in the ocean after recovery of pilot.
First 24-hour orbital flight.
Capsule's first orbital flight. Complete success.
Capsule makes a two-orbit, 3 hour 20 minute orbital flight. Complete success.
Complete success. Glenn spends 4 hours 55 minutes in space.
4 hour 56 minute orbital flight. Complete success.
3 day 22 hour orbital flight. First space 'rendezvous', with Vostok 4.
2 day 22 hour orbital flight. First space 'rendezvous', with Vostok 3.
9 hour 13 minute orbital flight. Complete success.
Record 22-orbit flight, 1 day 14 hours.
Record time in space, with 4 days 23 hours. Made 'rendezvous' with Vostok 4.
First woman in space. Made 'rendezvous' with Vostok 5. Flight lasted 2 days 23 hours.

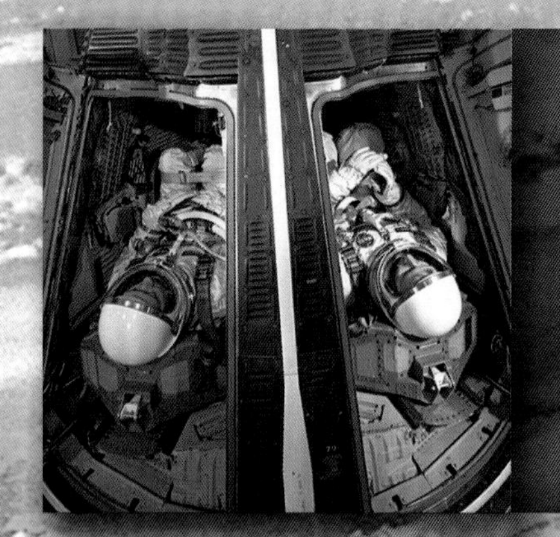

CHAPTER 2

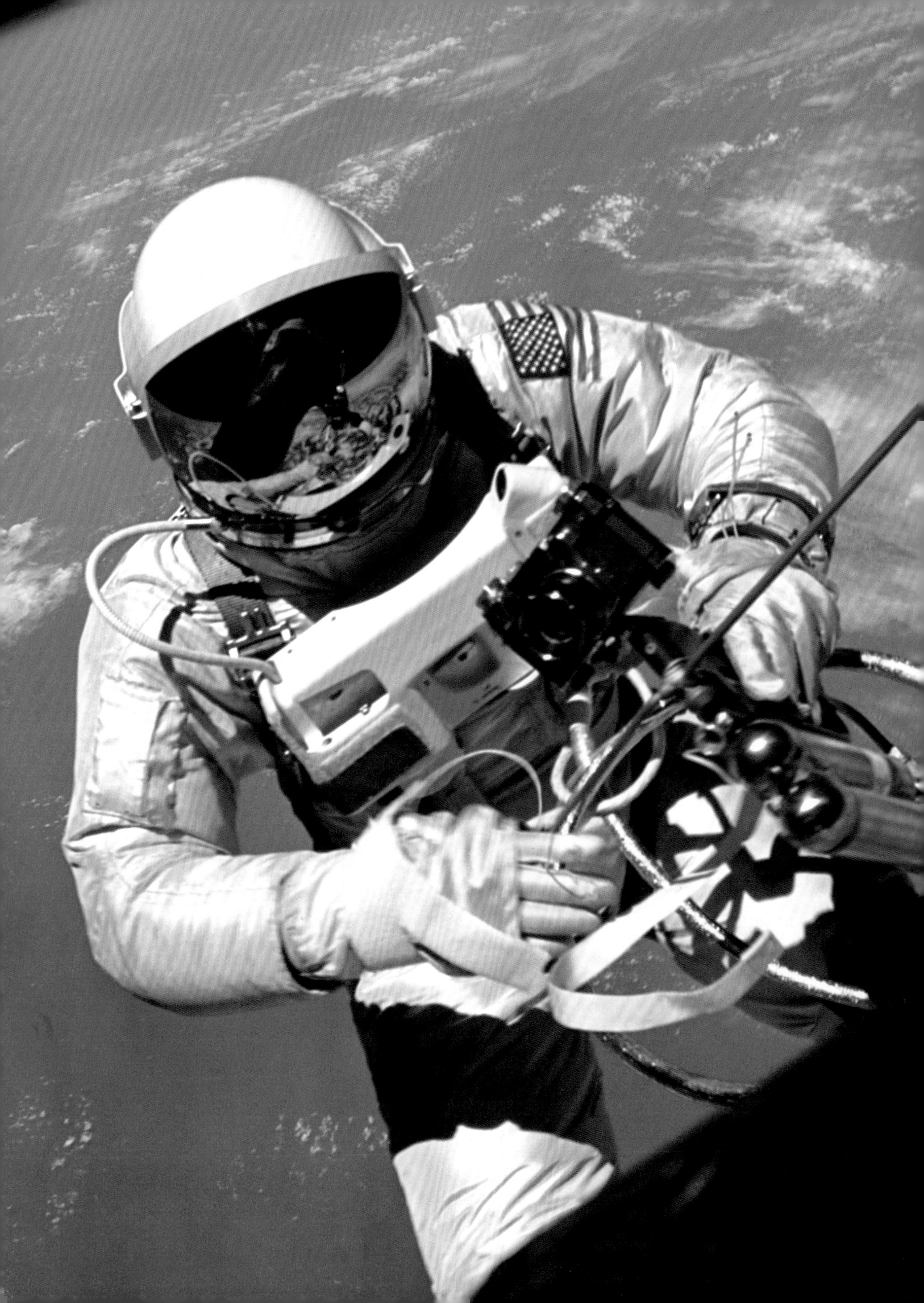

Rehearsal for the moon

The objective of the second phase of the American and Soviet space programmes was to send a crew of at least two men into space, evaluate the effects on them and try out extra-vehicular activities. In fact the Voskhod ('Sunrise') and Gemini programmes were expected to prepare the way for lunar missions, unlike the Vostok/Mercury projects, which were simply exploratory in nature. The Americans were rather sceptical. The launch of Sputnik 1 had been experienced as a kind of cosmic Pearl Harbor, and not only had the Russians taken a decisive lead in the intercontinental missile field, but they had scored further points by putting Gagarin into orbit. To reassure American public opinion, the new, young President John F. Kennedy made a celebrated declaration before Congress. The United States would, he said, send a man to the moon before the end of the decade.

While the patriotism of American citizens was certainly revived by this, NASA would have to work twice, perhaps three times, as hard. The young agency's budget was to see a huge increase. Yet again, though, it was the Soviets who scored first, passing close by the moon with their probes Luna 1 in 1959 and Luna 9 in 1966 (which actually landed on the lunar surface), and they were also the first to carry out a 'spacewalk'. However, these achievements must be tempered by the fact that the American Gemini programme gathered a great deal more information (it had several long spacewalks to its credit, as well as numerous manoeuvres and rendezvous).

Voskhod: two flights and it's over

The Voskhod programme's chief aim was to allow cosmonauts out of their craft wearing spacesuits in order to monitor their reactions and behaviour. The flights were also to be undertaken by at least two cosmonauts, in order to match the Americans, whose Gemini programme was based on a capsule carrying two men. From the purely technical standpoint, the Voskhod spacecraft was not very far removed from Vostok and would also form the basis of Soyuz. In fact, with the American announcement of the Gemini programme there had to be something to fill the gap between the end of the Vostok project (1963) and the entry into service of Soyuz (planned for 1965 at the earliest), and in order

Voskhod 2 was the first space mission in which an extra-vehicular activity (EVA) was carried out. The Voskhod craft was placed beneath the launcher's nose cone. This rocket was a version fitted with a third, more-powerful stage than the first Vostok 8K72, itself derived from the R-7. Designated 11A57, it could put a payload of 13,000lb into low orbit. Note the bulge on the nose cone created by the inflatable airlock.

To get out into space, it was decided to make use of an airlock made from a supple material which could be inflated automatically. When flat, the Volga airlock was only 27in in diameter by 30in high. It weighed 550lb with all its associated equipment and once deployed it measured 8.2ft, with an interior diameter of 3.2ft. The airlock was a vital piece of equipment, as Vostok's hatch had to be quickly closed or the depressurisation would have caused the on-board electronics to overheat.

to achieve this Korolev's OKB-1 decided to modify Vostok. The Voskhod spacecraft still comprised two modules, command and instrument; but a retro-rocket pack was added to the end of the command module, since, unlike its predecessor, Voskhod could not come down simply by atmospheric braking because of the higher orbits it was to adopt. However, unlike Gemini, Voskhod could not really manoeuvre in space (it could not change its orbit or carry out docking).

Three seats were planned inside the sphere. So as to fit the crew in comfortably, the designers not only had to abandon the idea of ejector seats, but also had to relocate the new seats at right angles to the position used in Vostok. Curiously, however, the instrument panel retained its original location, obliging the crew to crane their necks sideways to look at it. In a significant development from Vostok, Voskhod had a more sophisticated steering system, using an ion-flow sensor, so that adjustments could be made whether in daylight or eclipse. Voskhod was also heavier (5.2 tons) and benefited from a slightly more powerful launch vehicle in the shape of the 11A57, which replaced the 8K72.

The return to Earth was also quite different: since the crew had no means of escaping from the craft on lift-off, they were obliged to remain on board (which also meant that, for the first flight, the crew had no need for heavy and impractical spacesuits). A retro-rocket pack and the descent parachute shared the task of slowing Voskhod sufficiently to make a safe landing. A first mission without a crew was mounted from Baikonur on 10 June 1964. Then just two days later, with Vladimir M. Komarov as pilot, Konstantin P. Feoktistov, a scientist, and physicist Boris B. Yegerov on board, Voskhod 1 took off on 12 June at 07:26, local time. It was the first time in space history that a craft had carried more than one person. The mission was used to prove the spacecraft's worth and carry out a number of scientific experiments. Live television pictures were also transmitted, although some doubt remains as to their authenticity.

The Voskhod craft was fairly similar to Vostok, with its pronounced spherical shape. Note the cylinder on top of the sphere. This was the craft's retro-rocket system. The placing of the nosecone completed the assembly of the vehicle and the next step was to mount it on top of the launcher. The Vostok craft was initially called 'Vykhod', meaning 'exit', but this was thought to be too revealing of the mission's purpose.

The indispensable spacesuit

The Earth's gravity has conditioned all animal and vegetable life. Our atmosphere provides us with the air that is indispensable to our existence, but it also exerts a certain degree of pressure on us – around 14.7lb per square inch at sea level. A human being feels nothing, as our internal pressure keeps everything in balance. The air we breathe is generally composed of 78 per cent nitrogen and 21 per cent oxygen; the remaining one per cent comprises other gases. The higher one rises, the pressure decreases and the amount of oxygen diminishes. At an altitude of 18,000ft the atmosphere is half as dense as at ground level. Beyond 40,000ft, simply wearing an oxygen mask is not enough. Around 60,000ft, a suit is needed to restore ground level air pressure, as ambient pressure has become so low that the fluids in the human body are liable to boil. Clearly, extreme altitudes are harmful to man.

An astronaut therefore has to take his own environment with him if he is to go out into space or walk on a body such as the moon. Indeed, once in space, merely maintaining pressure and having an oxygen supply are not enough. Protection is also needed against the sun's rays, which can get warm very rapidly – there is no atmosphere to filter them – and quickly reach 120°C. By contrast, in the shade temperatures can drop to about –160°C. In addition meteorite particles might damage and pierce a spacesuit, which needs, therefore, to provide maximum protection. Finally, the suit must be properly ventilated to disperse the heat and sweat generated by the human body when it is moving. The design of the earliest suits was closely derived from those worn by high-altitude aircraft pilots. That worn by the Mercury astronauts, derived from the Navy's mark IV model, was a pressurised suit made by Goodrich, with a neoprene and nylon interior covered by a nylon and aluminium outer layer. Strictly speaking the suit was not fully pressurised, only becoming so in an emergency, which never happened.

For the Vostok spacecraft, the Soviets developed the Sokol SK-1, a high-performance, pressurised survival suit – since the pilot had to eject, he needed an independent oxygen supply. Coloured orange (to make it easy to spot), this suit was worn only by Gagarin, Titov and the other cosmonauts who took part in the first Vostok missions. It was succeeded by the Berkut suit, worn by Leonov for his first EVA (Extra-Vehicular Activity) in 1965. On this suit, the oxygen supply was not via an umbilical cord, but came from an independent system attached to the back. It could be operated outside the spacecraft for 45 minutes, which was sufficient for a simple spacewalk. There were two pressurisation systems: one kept the pressure at 0.27 bar and the other at 0.40 bar. Unfortunately the suit proved to be excessively rigid and Leonov, on his first spacewalk, experienced great difficulty in moving, and had to lower the internal pressure dangerously in order to achieve some flexibility.

The American Gemini missions as well as the first Apollo flights took place with a new piece of equipment made by the David Clark Company, which had managed to produce a particularly light, fully oxygen-pressurised suit that was much more flexible than those used on Mercury. Inspired by those worn by the X-15 pilots, these suits were designed not only for extreme depressurisation conditions or for spacewalks, but also for endurance flights. The G3C (used only on Gemini 3), the G4C (worn up to Gemini 7), and the G5C (used on Gemini 7) were made from six layers of nylon and nomex. Apart from the two oxygen connections, the chest had a single connection for the communications link and the medical sensors. The helmet was pressurised, but the high boots were of the standard type with nomex laces.

With all of them, the astronaut wore a special undergarment that made movement easier by eliminating direct contact between the skin and the suit, and forming a kind of second skin with the pressurised air. The G4C differed from the G3C by its additional white Mylar layer, to cope with the extremes of temperature in space. Cernan wore a variant of the G4C on Gemini 9. This suit's legs were covered in aluminium to give protection against the hot gases from the AMU rocket pack that it had been intended to use. It was on the 14-day Gemini 7 mission that the G5C made its appearance. A lighter version of the G4C, it hadn't been designed for an EVA, although perfectly adequate for use in the event of depressurisation. More comfortable, it could be removed during what were considered to be non-critical periods of the flight. This led to NASA, just like the Russians,

The Mercury mission suit, closely based on those worn by X-15 pilots.

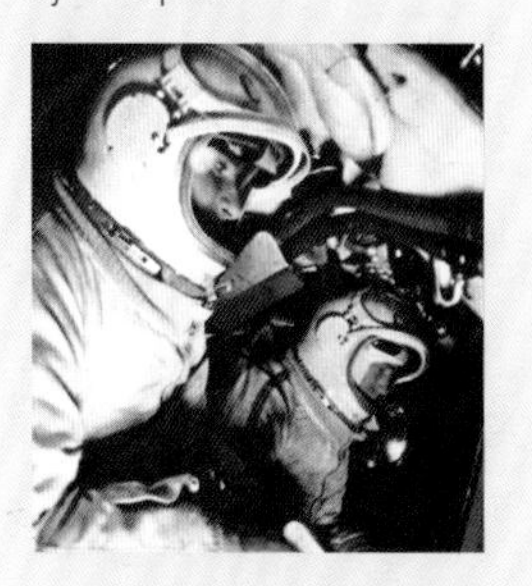

The Berkut suit, designed by the Soviets for spacewalks. Partly self-contained, with the oxygen carried in the backpack, this suit proved to be too rigid when pressurised and was abandoned.

Stafford and Schirra (Gemini 6A) are wearing the G4C suit, which was the most frequently worn suit during the Gemini missions. This type of suit was worn by Ed White when he carried out the first American EVA.

A completely new suit, the A7L, was developed for the Apollo missions. The 'goldfish-bowl' helmet is clearly visible here. It was partially covered to protect the head from the sun's rays. This suit was fully pressurised and incorporated a self-contained oxygen and electricity supply.

The Soviets developed the Orlan suit during the Salyut programme. This self-contained suit was derived from the Krechet model intended for survival on the surface of the moon. Orlan, still in use, is a one-piece suit, including the helmet. The cosmonaut gets into it from the back by opening the backpack like a car door.

For missions not involving an EVA, the Russians have used the Sokol SK-1 since 1971. It has been worn since the tragic accident to Soyuz 11 in 1971.

The EMU outfit is the suit currently used by NASA for spacewalks.

adopting a simpler, more comfortable suit for the Apollo missions, reserving the pressure suit for lift-off, re-entry and EVAs.

The early Apollo missions were devoted to equipment testing. As there were no EVAs planned, NASA made use of a modified version of the G4C designated A1C. However, the Apollo 1 accident in 1967 meant that the suit could not be evaluated and from Apollo 7 onwards the new A7L suit, designed by ILC Dover, was tested. This suit was to become standard equipment throughout the whole programme. Quite substantial (160lb) and somewhat complex (putting it on was a lengthy and tedious process, even with the lighter A7LB versions), it was a veritable suit of armour with metal joints where the pressurised, ball-bearing-mounted gloves were attached, integral boots, the well-known 'goldfish-bowl' helmet with panoramic vision, and an external fire-resistant layer. Each suit was made-to-measure and each astronaut had three of them. The astronauts who walked on the surface of the moon wore a different suit from the CMP (the Command Module Pilot), who remained on board the CSM in orbit.

Their spacesuits had six connections (instead of three) and no fewer than 21 insulating layers. Next to his skin, the astronaut wore a water-cooled garment (the Liquid Cooling Garment or LCG), preventing excessive sweating and eliminating the problem of fogging of the helmet visors. Although this under-suit was worn all the time, it functioned only when the crewmember left the spacecraft. In addition, the 'moonwalker' would wear a very light, aluminium-film-covered outfit next to the spacesuit. To go out onto the lunar surface, he also had to put on an over-garment with a hood woven in aluminised Mylar to protect him from the sun's rays. Over-boots made from the same material were also provided. The A7L-B suit was worn by the crew on Apollo missions 15, 16 and 17. It was less rigid – thanks to a joint at waist level – so that the astronaut could sit on and get out of the Lunar Rover.

For their part, the Russian firm of Zvezda developed the Yastreb suit from 1965 onwards. This was more rigid to counter the 'ballooning' effect of the Berkut suit and was notable for its curious collection of cables and pulleys designed to give the forearm flexibility. In the end, the suit was little used (for Soyuz 4 and 5 only, as docking failures on Soyuz 7 and 8 caused the planned EVAs to be aborted). The Krechet suit was to have been worn by Soviet cosmonauts on the moon. It was an original design, being made in one piece, including the fixed helmet and the independent survival system. The wearer got into it via a 'door' in the back. His legs went in first, followed by his torso, head and arms; then a fellow crewmember would close it up. There was, therefore, no external flexible joint and the suit's integral design was very practical. Although the moon programme was eventually abandoned, the Krechet suit still found a use as the basis of the Orlan suit, still in service for EVAs. Used in the Salyut, Mir and ISS missions, the Orlan has undergone various modifications (it is now entirely self-contained, whereas the original models had an umbilical cord for the oxygen supply).

NASA began to think about a new self-contained spacesuit in 1974 with spacewalks from the Shuttle in mind. Adopted in 1984, the EMU (Extra-vehicular Mobility Unit) was more than just a suit: it was a complete survival, communications and work system. Made by Hamilton Sunstrand and ILC Dover, it consisted of an upper part (Hard Upper Torso, or HUT) comprising the torso with all the survival and electric systems, the arms, gloves and helmet; and a lower part (Lower Torso Unit, or LTU), with the lower body, legs and boots. These two parts were then locked together. As with the A7L, the astronaut first put on a cooling undergarment (Liquid Cooling Ventilation Garment, or LCVG). Taking longer to put on than the Orlan garment, which was very quick, the EMU suit kept the same helmet as for the Apollo missions, but now fitted with a camera and integral lights. Apart from the gloves, which were made-to-measure, the EMU was made using several different sizes of HUT and LTU sections to allow for it being worn a number of times by different crewmembers.

The Soviets (and later the Russians and all the cosmonauts launched from Baikonur) regularly wore a Sokol-K1-type suit. This was only worn during 'transport' flights, to ensure the crew's safety after the tragic accident with Soyuz 11. NASA chose the same solution with the ACES (Advanced Crew Escape Suit), regularly worn by Shuttle crews since the *Challenger* accident. A vivid orange colour, this suit came from the David Clarke Company.

A memorable walk!

The Voskhod 2 craft, which lifted off on 18 March 1965, was slightly different from its predecessor (it was the 3KD version, not the preceding 3KV). An inflatable decompression airlock named 'Volga' was sited facing the entrance airlock, allowing a crewman to exit into space. This 88cu ft airtight inflatable compartment permitted a man to get outside in complete safety, since it was not possible to depressurise the entire spacecraft because the flight and monitoring instruments were cooled by the internal ventilation system. The mission was planned for only two men, as they were going to be wearing the cumbersome Berkut suit and there would have been insufficient room for three suited cosmonauts in the Voskhod's confined space. After a minor scare during the launch – an alarm went off for no known reason – it fell to cosmonaut Alexei Leonov to make the first spacewalk, with the second crew member, Pavel Belyayev, remaining on board.

Once Voskhod 2 was in orbit, the two men made their preparations. Leonov had to put on an extra piece of equipment that supplied oxygen to his Berkut suit and would provide him with a maximum of 45 minutes of air. The Volga inflatable compartment was deployed and pressurised – a process that took seven minutes – then Leonov entered while Belyayev closed the airlock behind him. When the seals had been checked for airtightness, the unit was depressurised, and Leonov opened the overhead panel and went out into space. An umbilical cord over 50ft long kept him secure, but he recalls being restricted in his movements, partly by the cord and also by his suit, which had been overinflated by the air supply. Cramped in his Berkut, Leonov was unable to take any photographs – he simply couldn't reach the shutter release – and the only known pictures were taken by the automatic cameras mounted near the airlock. After 12 minutes Leonov came back on board without mishap and Belyayev jettisoned the Volga. The spacecraft returned to Earth after 17 orbits, more than 24 hours after launch.

Several problems arose during the flight: Leonov had considerable difficulty getting back into the airlock because of his overinflated suit and he was obliged to release some air to regain normal size. This was a very risky process for the cosmonaut, whose body temperature and heartbeat rose rapidly. The

Leonov made the first walk in space. He joined the cosmonaut group in March 1960, at the same time as the other men and women of the Vostok programme. It was originally intended that the Vostok 7 to 10 missions would carry on throughout 1964. While the future missions were in preparation, Leonov finished his training for his next flight on Vostok, but with the programme having been dropped he was assigned to the Voskhod 2 mission and became the first man to 'walk' in space.

A Gemini craft is unloaded from a Douglas C-124 Globemaster of the MATS (Military Air Transport Service). The re-entry module (the dark part under the sheeting) was delivered with the equipment module (the rear, white part), which was the heart of Gemini. This was the powerhouse of the module, containing the electricity generator, water and oxygen reserves. The cylindrical nose containing the radar has not yet been fitted.

return to the spacecraft was eventually achieved safely and Belyayev took over manual control of the retro-rockets. Once again the two modules did not separate before re-entry into the atmosphere, causing unintended spinning. Finally, the craft landed in a wooded area deep in the Urals and the crew spent a night surrounded by wolves before they were found!

Much was learned from these experiences, but later in 1965 the Soviets decided to halt the Voskhod programme (perhaps because it was proving to be too risky) and to concentrate instead on reaching the moon with Soyuz, which entailed a radical change of direction. However, several other Voskhod flights (3 to 6) had been planned for the first half of 1965 with periods in space ranging from 12 to 15 days (with just one crew member). But while the spacecraft needed significant modification, another issue had arisen with the choice of cosmonauts, which had become the subject of much political discussion – the men approached by Kamanin were consistently rejected by the party brains: this was the case

Models of the Titan rocket and the Gemini craft are tested in the wind tunnel at Langley, which was the first American civil aeronautics laboratory. Established in 1917, it was first taken into NACA, and then NASA, and is NASA's oldest facility. It has more than 40 different types of wind tunnel. With the Gemini programme, Langley became the training base for orbital rendezvous simulations and for testing lunar vehicles and spacesuits.

One of the most amazing projects in the Gemini programme was the plan for Gemini to land at Edwards on its return to Earth using an inflatable delta wing. This was one of the reasons why ejector seats were installed. The idea came from the engineer Rogallo, who invented the modern delta wing. It may seem a bit hare-brained, but the tests proved fairly positive, even though the deployment of the wing was causing problems (the reason why it was eventually abandoned). The inflatable wing would have been very cheap and it would have obviated the need to mobilise a large naval fleet every time the capsule had to be recovered. For Gemini's first manned flight, no fewer than 11,000 men, 126 planes and 27 ships were mobilised!

Gemini was the only manned American spacecraft to be fitted with ejector seats (the Shuttle was to have them, but only for the test flights). As a piloted return using the inflatable delta wing had been under study, ejector seats would have been needed should a problem have arisen. The seats also allowed the craft to be evacuated at any point between ignition and 65,000ft. The two hatches would have opened and the seats fired out of the craft. In addition, the seats were mounted on runners, which greatly eased access to the craft. Before lift-off the astronaut got into his seat (with the capsule vertical), and was strapped in and then slid back into the capsule. Weber Aircraft of Burbank designed and built the ejector seats, which were never used.

with Volynov, because he was Jewish, and with Katys, because his father had been executed in one of Stalin's purges!

In the meantime the Gemini programme was proceeding briskly. Gemini 5 had brought the record for a manned flight up to eight days, while Gemini 7 increased it to 14. It was therefore more or less essential for the Soviets to aim for a flight of at least 20 days. But poorly defined, neglected by Brezhnev the new boss in the Kremlin, and held up by the death of Korolev in 1966, the Soviet space programme was in crisis. Voskhod 3's flight was postponed to1966 and was then forgotten through lack of development and a will to carry it through.

The American Gemini programme

The race to the moon being well under way in Uncle Sam's country, it was vital to plan for the future Apollo missions. With the Gemini programme, American astronauts were to undertake, evaluate and test long-duration flights. The USAF also took an interest in the programme, which took two directions: the MOL (Manned Orbiting Laboratory) was to be the early stage of a manned observation station; and project Blue Gemini, which was a military programme intended to develop a Gemini module that could rendezvous with 'enemy' satellites. The USAF was responsible for a portion of the funding and the MOL project took shape with the launch of an unmanned Gemini capsule – in fact Gemini 2 was reused, this being the only occasion that a spacecraft was reused before the Shuttle era. However, the USAF would not allow the Navy to have the responsibility of recovering the capsules, and this internal wrangling – combined with the development of the lunar programme, now given priority – brought an end to these tests before the close of the decade.

The second American manned programme was initially designated 'Mercury II', and finally 'Gemini' for the simple reason that there were always two crewmembers. It was also the name of the third constellation of the zodiac, and was a name clearly evocative of space conquest. A lot has been written on the supposed benefits of the Gemini programme, which was considered a waste of time by some experts who would have preferred to put everything into the lunar programme. But the Apollo lunar programme was already up and running (it had been started just after Kennedy's speech), and since its development would take some time the Gemini programme provided an opportunity to put into early practice some of the technology that would be useful in Apollo. The experiences of weightlessness and manoeuvring in space were also profitable. In the event, the Gemini programme allowed NASA to gain the upper hand over the Soviets and to demonstrate its capability in long-duration flights.

Anatomy of Gemini

McDonnell-Douglas's design for the Gemini capsule followed the same form as the one built for the Mercury project, but it was larger and had different design features, being able to manoeuvre in any direction and change its orbit. It is often forgotten that the Gemini project was technologically more up-to-date than Apollo, as the latter had been started two years before Gemini and had not benefited from the latest innovations in aeronautical technology. The Gemini spacecraft consisted of three modules, although often considered as two, as the second and third were actually part of the same system.

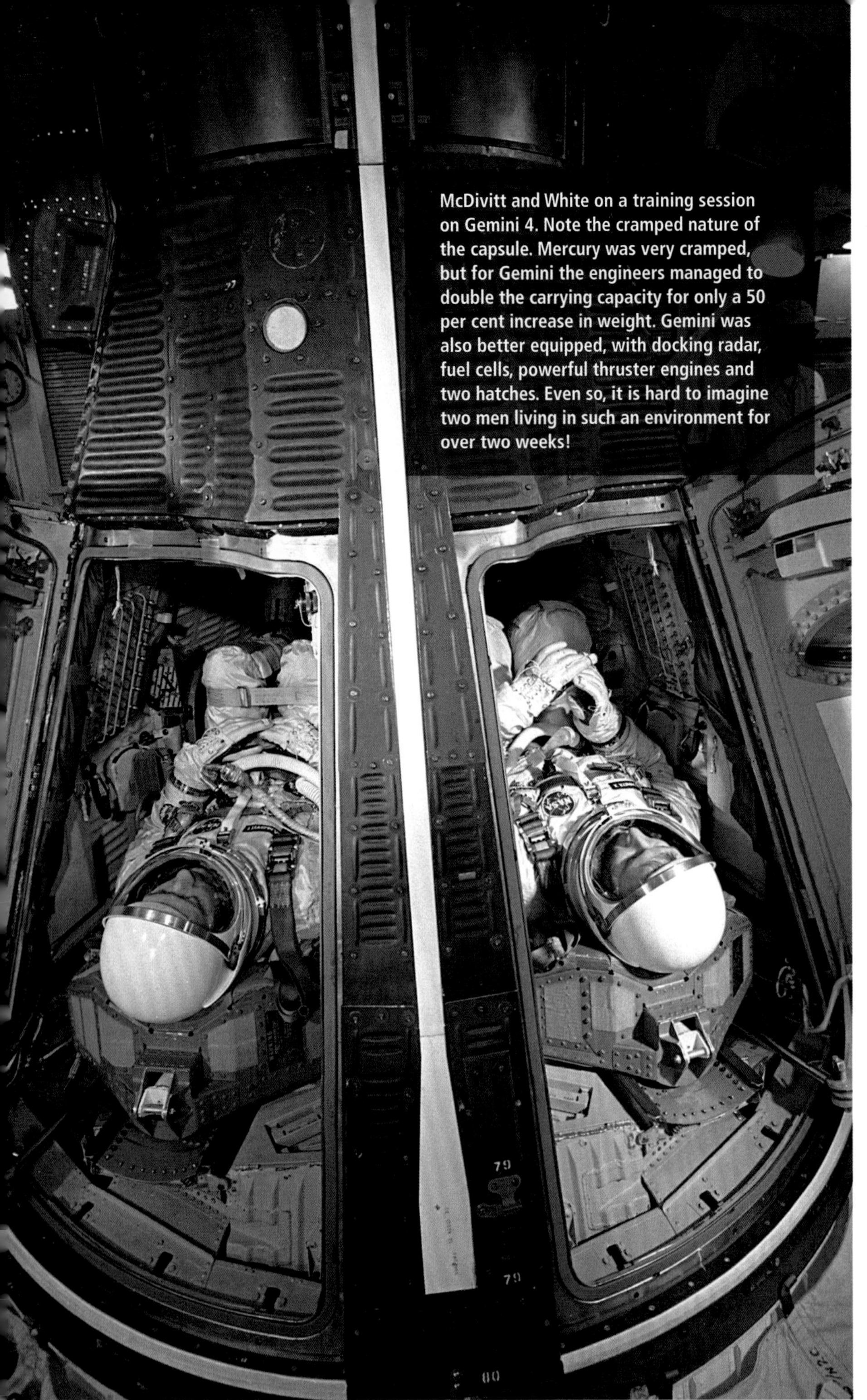

McDivitt and White on a training session on Gemini 4. Note the cramped nature of the capsule. Mercury was very cramped, but for Gemini the engineers managed to double the carrying capacity for only a 50 per cent increase in weight. Gemini was also better equipped, with docking radar, fuel cells, powerful thruster engines and two hatches. Even so, it is hard to imagine two men living in such an environment for over two weeks!

The first was the Re-entry Module (RM), 11ft long, which included the radar nose for rendezvous in space, the descent parachute, the navigation instruments and the astronauts' living compartment with its ejector seats. The latter had been preferred to the Mercury escape tower, as the Titan II rocket used in launching Gemini had a negligible risk of exploding on the launch pad. Gemini had also been designed for space rendezvous with the Agena module (a highly modular upper stage mounted on a Thor, Atlas or Titan rocket), capable not only of putting satellites into orbit but also of being used as a rendezvous target during the programme. The Gemini capsule was fitted with radar and an inertial navigation system and made much use of very lightweight materials in its construction. Spacewalks were made easier by the large openings with proper doors, located in front of the astronauts. The other side of the coin, however, was the cramped space, with the two men being literally elbow to elbow in very restricted surroundings, which made long missions particularly burdensome.

The second module was the AM (Adaptor Module), only 3ft long and carrying the retro-rockets. The third module, measuring 4.6ft, was the EM (Equipment Module), where the energy supplies (water tanks, oxygen, cooling pumps, electricity) were located, along with the rockets for adjusting the orbit. Curiously, it had originally been intended that Gemini would not ditch in the sea using parachutes on its return to Earth, but would instead be piloted using flexible wings, derived from research undertaken by Francis Rogallo (the father of the hang-glider), and land on skids on the salt flats at Edwards. However, work on the wings became so delayed that the idea

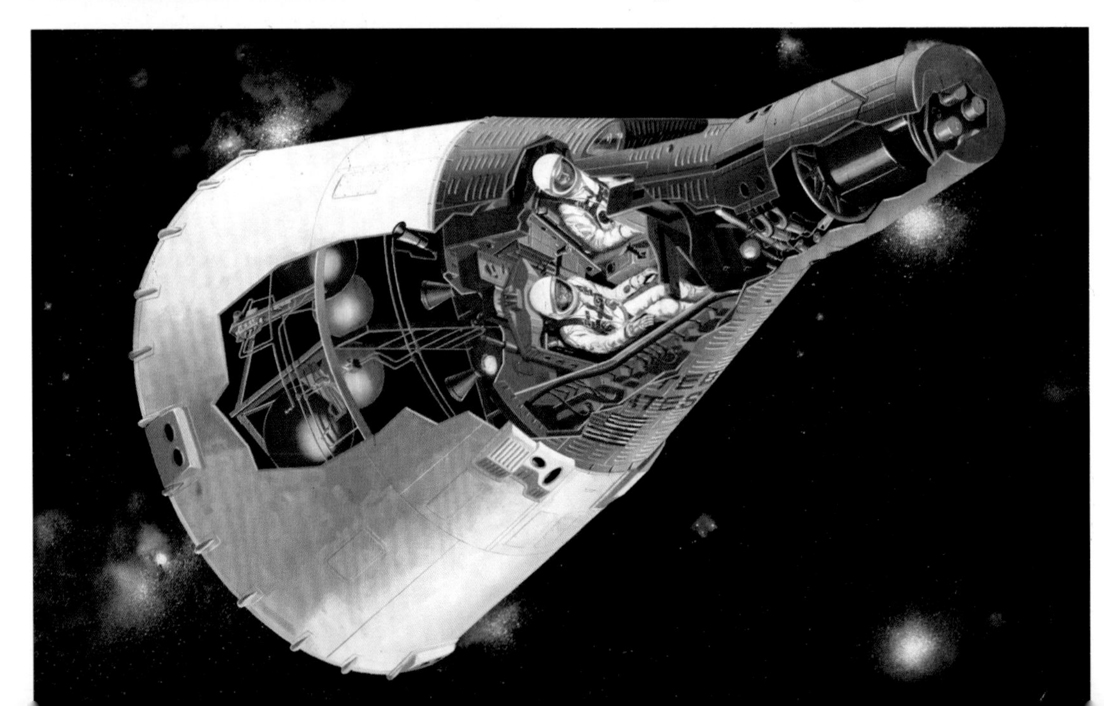

This exploded view reveals Gemini's interior with the RM (Re-entry Module) and the two crew members strapped in to their ejector seats facing the hatches. Every available space was accounted for. At the base of the cylindrical nose can be seen the propergol-powered thrusters, used on re-entry. Behind the heat shield is the AM (Adaptor Module) with the engines controlling pitch, yaw and manoeuvring (on the perimeter) as well as the retro-rockets (recognisable from their red tank). The last part is the EM (Equipment Module) with the propergol tanks (the large red spheres) and the oxygen and water tanks (yellow). The thin tubes supply the engines controlling yaw. The cooling system – not visible here – was also installed in the EM.

Life support for 14 days in space

The NASA-McDonnell Project Gemini is the major link between Project Mercury and Project Apollo (this nation's first flight to the moon). It will give our space effort vital information on prolonged spaceflight effects and will also be used to test space rendezvous techniques.

Gemini's advanced environmental system will keep the spacecraft's two astronauts comfortable for two weeks of continuous orbital flight. Garrett-AiResearch builds the system that provides a breathable atmosphere, pressurization, temperature control, ventilation and atmosphere purification in the two-man spacecraft and in both astronauts' suits for the entire flight. AiResearch also supplies the supercritical cryogenic oxygen and hydrogen tankage system for the fuel cell power supply.

This major contribution to the advancement of space travel is one more example of Garrett's proved capability in the design and production of vital systems and their components for man's most challenging exploration.

THE GARRETT CORPORATION • AiResearch Manufacturing Divisions • Los Angeles 9, California • Phoenix, Arizona • other divisions and subsidiaries: Airsupply-Aero Engineering • AiResearch Aviation Service • Garrett Supply • Air Cruisers • AiResearch Industrial • Garrett Manufacturing Limited • Garrett International S. A. • Garrett (Japan) Limited

IME, MARCH 15, 1963 10

was dropped. Furthermore, the Gemini project was sufficiently advanced to consider it being used for a trip to the moon, at a much lower cost than that of Apollo (the Gemini project cost 'only' five per cent as much as the Apollo programme), but its lack of creature comforts was too big a handicap. The main launch vehicle was a Titan rocket in its mark II version, derived from an intercontinental ballistic missile. Because of the rocket's military origin all the launches took place from the US Air Force LC 19 pad at Cape Canaveral.

Long-duration missions

As usual, NASA proceeded cautiously: the first two flights, Gemini 1 and Gemini 2, were unmanned test missions. Gemini 1 was deliberately destroyed on re-entry on 8 April 1964, but Gemini 2 returned safely after a flight lasting less than 20 minutes on 19 January 1965. These flights served to test not only the capsule itself, but also the Titan II rocket. After this the serious business began with Gemini 3, which carried Virgil I. Grissom and John W. Young on 23 March 1965. This four-hour 52-minute flight was intended to test the ability of the crew to change their orbit, which was duly carried out an hour and a half after lift-off, Gemini moving from 75 miles' altitude to 110 miles at a speed of 50ft/sec. Everything went normally but for one tragi-comic incident: Young had managed to smuggle aboard a sandwich that he offered to Grissom, who couldn't stand freeze-dried food. In the cabin's weightless environment, however, the sandwich started to come apart, which could have had disastrous consequences if crumbs had got stuck in any of the electronic systems, particularly the ventilation. The two men had great difficulty in retrieving all the pieces and Young was severely reprimanded by NASA…and by Grissom, who complained that he'd forgotten the mustard!

Gemini 3 was also the last spacecraft to receive a nickname. Grissom came up with 'Molly Brown', a light-hearted reference to *The Unsinkable Molly Brown*, a Broadway hit at the time. Grissom's Mercury capsule had, of

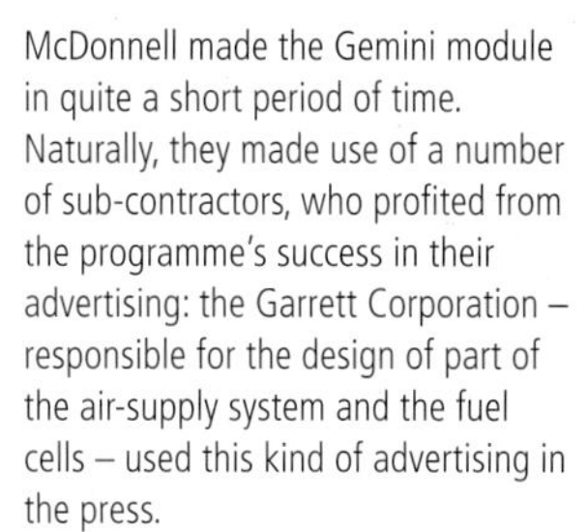

McDonnell made the Gemini module in quite a short period of time. Naturally, they made use of a number of sub-contractors, who profited from the programme's success in their advertising: the Garrett Corporation – responsible for the design of part of the air-supply system and the fuel cells – used this kind of advertising in the press.

Grissom and Young took off on 23 March 1965. It was a mission of firsts: the first manned Gemini flight, the first time that an astronaut had flown twice and the first time two Americans had flown together in space. The flight lasted for only three orbits. The Titan rocket's lift-off from pad 19 was delayed for 35 minutes because of a technical problem, although this did allow the sky to clear. The lift-off was a model of smoothness, but there were some technical issues that were quickly resolved thanks to the many hours of training that had been done. The return was less positive, with Gemini 3 coming down over 50 miles away from its intended point of impact. Furthermore, during the parachute descent the capsule jerked violently and the two men were thrown forward, cracking the front of Grissom's helmet. They remained floating in the capsule for 30 minutes and when the Navy frogmen finally opened the hatches, the crew needed no persuading to evacuate 'Molly Brown'!

The Apollo 4 mission started on 3 June 1964. Ed White and James McDivitt arrived at the launch pad over an hour and a half before take-off. They were taken up to the Gemini 4 capsule in the lift. The two men were assisted by McDonnell personnel, but getting into the craft was relatively simple. White (lying down on the right) nevertheless had to make use of the spacesuit's ventilation system because his sweat was causing his helmet visor to mist up. He was the only one to carry an extra helmet, protected inside a cover. The visor had a gold film designed to give protection from the sun's rays when outside the craft.

course, sunk rapidly after splashdown and he hoped to cock a snook at this disastrous incident. When he was asked to think of another name, he picked out 'Titanic'. Eventually, the NASA directors accepted 'Molly Brown', but it was the last time a spacecraft was named until the Apollo 9 mission, where there was a need to distinguish the capsule from the moon-lander. Gemini 4 was a landmark in the American conquest of space, for several reasons. It was the first time a flight was broadcast live to the United States and Europe; it was the first time that the Houston control centre was used; it was the first

The tower is about to be lowered and everything is on course as the Titan rocket lifts off at 10:16 local time. Parts of the launch pad are sprayed with water to keep down the temperature around the tower. It is a perfect lift-off and the stages separate one by one. The spectacle was broadcast live by the American TV networks, and around 1,100 journalists from a variety of media covered the event.

In low orbit, Gemini 4 is about to begin its approach to the final stage of the Titan II rocket that is to serve as the target for a virtual docking. But the many attempts to get close fail and the crew is forced to abandon this first part of their mission.

Over the Indian Ocean, after an exhaustive checklist, the two men depressurise the capsule and White opens his hatch. Carefully, he extricates himself, installs a camera and starts to float just above Gemini 4. Using his gas-gun he moves about easily. He will remain outside for just 15 minutes. He returns to the craft without any difficulty, removes the camera and hands it to McDivitt. But closing the hatch proves more difficult than imagined, despite the assistance of his colleague.

American flight lasting more than 24 hours; and finally, it was the first time an American had done a spacewalk, or EVA (Extra-Vehicular Activity). Edward White, assisted by James McDivitt, spent 22 minutes out in space, moving around a few metres from the capsule (to which he was linked by an umbilical cord) with the aid of an oxygen powered-gun. After about 16 minutes he was given the order to return inside the capsule: 'It's the saddest moment of my life!' he said. There were no problems getting back on board, but the hatch proved difficult to reseal.

While the EVA had been intended to show that the Americans could do just as well as the Soviets, the principal objective of the mission was to 'fly' in formation with the second stage of the Titan II rocket floating in space. This was done to try out the rendezvous procedure, which would become common practice in the future. It was not a success, as the Gemini capsule used up too much fuel in its approach to the target. However, the NASA experts drew a number of conclusions from the mission: it was better to put a spacecraft into a lower orbit than its rendezvous target, the latter thereby having greater speed, allowing a quicker transition into a higher orbit afterwards. Several experiments were carried out on board to observe human performance in space, both during sleep and while awake. Many very good photographs were taken, but on the 48th orbit the main IBM computer broke down. This was bad publicity for the company, which had taken a centre spread in the *Wall Street Journal* saying that its computers were so reliable that they had been chosen by NASA!

On the 72nd orbit, Gemini 4 started its re-entry in manual mode. Everything went successfully, although splashdown was 50 miles from the intended spot.

After 97 hours and 28 minutes, Gemini 4 began its re-entry into the atmosphere…in manual mode, thanks to a computer fault. The equipment module was jettisoned and the two astronauts watched as it turned red, caught fire and was burnt up. The retro-rockets came into action and the long descent began. At around 40,000ft the braking parachute deployed, followed at 10,000ft by the main parachute. This opened fully in one action, but to slow down the capsule so that it hit the water as gently as possible – Mercury floated upright, but Gemini lay horizontally – additional webbing was deployed at about 5,000ft to turn the capsule onto its side. Gemini 4 came down further on than intended, but the crew was soon joined by an armada of ships. As with all the American capsules, a fluorescent green dye was used to colour the surface of the water. The crew was soon on board the American carrier *Wasp*, from where they were able to speak to President Johnson.

Technical details

NASA designation: Titan II

Type: orbital launcher

Country: United States

Manufacturer: Martin

First launch: 16 March 1962

Last launch: 18 October 2003

Power on lift-off: 844 tons

Engines:
– First stage:
2 Aerojet LR-87 rocket engines running on an aerozine 50/hydrogen peroxide mixture giving 1,900kN of thrust for 139 seconds

– Second stage:
1 Aerojet LR-91 rocket engine running on an aerozine 50/hydrogen peroxide mixture giving 1,900kN of thrust for 180 seconds

Height: 103ft

Payload: 6,800lb

Total weight: 151.5 tons

The Titan launcher: a big little rocket

The Titan rocket constitutes a very large family of launchers, with the latest derivatives still in service after a half-century of existence. The first, Titan I, was developed by the firm of Martin in the mid-1950s as an intercontinental missile. It was in competition with the Atlas missile, but the Air Force wisely chose to support the development of both in case the Atlas should prove ineffective. Given the designation SM-68, the Titan missile used the same fuel mixture as the Atlas, namely liquid oxygen and kerosene. The engine was certainly different, but while Atlas was distinguished by its very light, pressurised structure, Martin opted for a more solid, classic design which necessitated the adoption of two stages.

The Titan II rocket that succeeded it was more sophisticated. As a missile, it could be put into action more rapidly: the larger tanks could be filled in advance and the fuel could be stored at ambient temperatures thanks to the use of a mixture of aerozine 50 (fuel) and hydrogen peroxide (oxidant). Titan II could also carry a more powerful nuclear warhead. It entered service in 1963 under the designation of LGM-25C and became the most reliable and powerful rocket of its time, notably because of its W-53 nuclear warhead with inertial guidance. Very soon the Air Force and NASA alike realised its potential. Thus, the whole of the Gemini programme made use of this rocket, which continued its successful military career until 1987. By this time the missiles were relegated to lesser duties, being used for civil and military launches and known as Titan 23B and 23G. In the meantime the Air Force had developed the Titan III rocket, identifiable by its two lateral boosters. It was used to put military satellites into orbit, but once again NASA also made intensive use of it, notably for the Voyager probes.

Technical details

Name: Voskhod 11A57

Type: orbital launcher

Country: Soviet Union

Manufacturer: Korolev

First launch: 16 November 1963

Last launch: 29 June 1976

Power on lift-off: 3,999.93kN

Engines:
– First stage (0):
4 RD-107 rocket engines running on a liquid oxygen/kerosene mixture giving 98 tons of thrust for 119 seconds

– First stage (1):
1 RD-108 rocket engine running on a liquid oxygen/kerosene mixture giving 91.5 tons of thrust for 301 seconds

– Second stage (2):
1 RD-0108 rocket engine running on a liquid oxygen/kerosene mixture giving 28.9 tons of thrust for 240 seconds

Height: 115.5ft

Payload: 13,000lb

Total weight: 293.7 tons

The Voskhod launcher: a development in the right direction

The Voskhod was an R-7 rocket, almost identical to the one that had served successively in the Sputnik and Vostok programmes. It retained the four engines (with four combustion chambers) of the 8K72 rocket that formed the first stage, distributed around a central engine with four combustion chambers (the second stage). The difference came from the imposing third stage with its considerably more-powerful engine (but which, empty, was little heavier than the 8K72's old engine). Designated 11A57 in the Soviet classification system, the Voskhod rocket could put heavier loads into low orbit. In fact, the third-stage RD-0180, with its four combustion chambers, had originally been designed for the 8K78 rocket intended to launch interplanetary probes and Zenit-type spy satellites. After launching Voskhod craft, the 11A57 was heavily used in launching the Cosmos and Zenit military spy satellites.

A view of Cape Canaveral in Florida, seen from Gemini 5. The crew found this long mission rather tedious, but it gave the opportunity to show that human beings suffered no undue stress from being in orbit. This was an important step, as the moon missions were also to last about eight days.

An Ethiopian postcard from September 1965, honouring the two American astronauts, Cooper and Conrad. True ambassadors for the United States, they travelled all over the world. Their Russian counterparts did the same in pro-Soviet countries.

Records smashed!

With Gemini 5, the Americans were finally able to exceed Soviet achievements, as well as to carry out a rehearsal of the flight duration required for a journey to the moon and back. Gemini 5 and its crew, Pete Conrad and Gordon Cooper, would actually spend almost eight days in space. The capsule and its launcher lifted off on 21 August 1965. Gemini 5 was identical to its predecessors, but its fuel cells were of a new type specially designed for lengthy missions.

The crew were to have practised a space rendezvous, but electrical problems prevented this and other tasks from being carried out. Instead they had to make do with a 'virtual' rendezvous, targeting a particular point in space and then manoeuvring with the help of OAMS (Orbit Attitude and Manoeuvring System). After the successful execution of this manoeuvre on the third day, however, Conrad and Cooper began to feel the effects of boredom, all the more so as one of the manoeuvring rockets failed, restricting their movement. Nevertheless, the mission was very important, as the eight days in orbit represented exactly the time needed for a return trip to the moon. From a medical perspective, the crew complained of suffering from cold and were bothered by the stars continuously shining in through the window, which they had to block out.

The return to Earth was controlled entirely by the on-board computer, which functioned perfectly, but a human programming error caused Gemini 5 to splash down about 80 miles short of the intended spot. On a more trivial note, Gemini 5 was the first NASA mission to have its own logo – a pioneers' covered wagon.

Conrad and Cooper set off to join Gemini 5. Recognisable behind them is Joe Schmidt, the American spacesuit man. The astronauts are carrying their own breathing system, which they will disconnect before plugging themselves into the capsule's system.

Brought by helicopter to the carrier *Lake Champlain*, Conrad and Cooper look happy! They have just spent over eight days in Gemini 5. A beard is to be expected, but the Gemini missions brought to light a little-discussed problem: hygiene. The crew wore their pressurised suits 24 hours a day. For the Apollo missions a new range of more-comfortable and hygienic clothing had to be designed.

The photographs taken during White's EVA were of excellent quality. They were taken using a Zeiss Nikon camera used by the American Air Force.

The two crews of Gemini 6A and Gemini 7 with the pairing of Stafford-Schirra and Lovell-Borman. Stafford was a fighter pilot before becoming a highly qualified instructor. He joined the astronaut group in 1962 to take part in the Gemini programme and participated in both Gemini 6A and Gemini 9. He was the recognised space-rendezvous specialist, which earned him a place leading the Apollo 10 mission and later Apollo-Soyuz. Schirra had an unusual family background: his father, an aviation enthusiast, had been one of those 'magnificent men in their flying machines' and his mother had been an aviation stuntwoman who performed acrobatics on aircraft wings. Such barnstorming shows, as they were called (because many of the aerobatic flyers amused themselves by flying through barns), were very popular in the United States during the interwar period. Schirra later served in Korea, and then became a test pilot. He participated in the Sidewinder air-to-air missile programme and the entry into service of the Douglas F-4 Phantom, before becoming one of the Mercury seven. He flew with Gemini 6A and later on Apollo. Jim Lovell is better known for his role in Apollo 13 than for his participation in Gemini 7. A test pilot at the Naval Air Test Center at Patuxent River, he was to have been one of the seven men on the Mercury programme, but was dropped because of some minor medical complications. Eventually selected for the second group in 1962, he took part in two Gemini missions (Gemini 7 and Gemini 12), replaced Michael Collins on Apollo 9 and then took charge of the ill-starred Apollo 13 mission. Frank Borman started his career as an astronaut with Gemini 7. A pilot from a very young age, he joined the Air Force at the age of 22 and was appointed commander of Apollo 8.

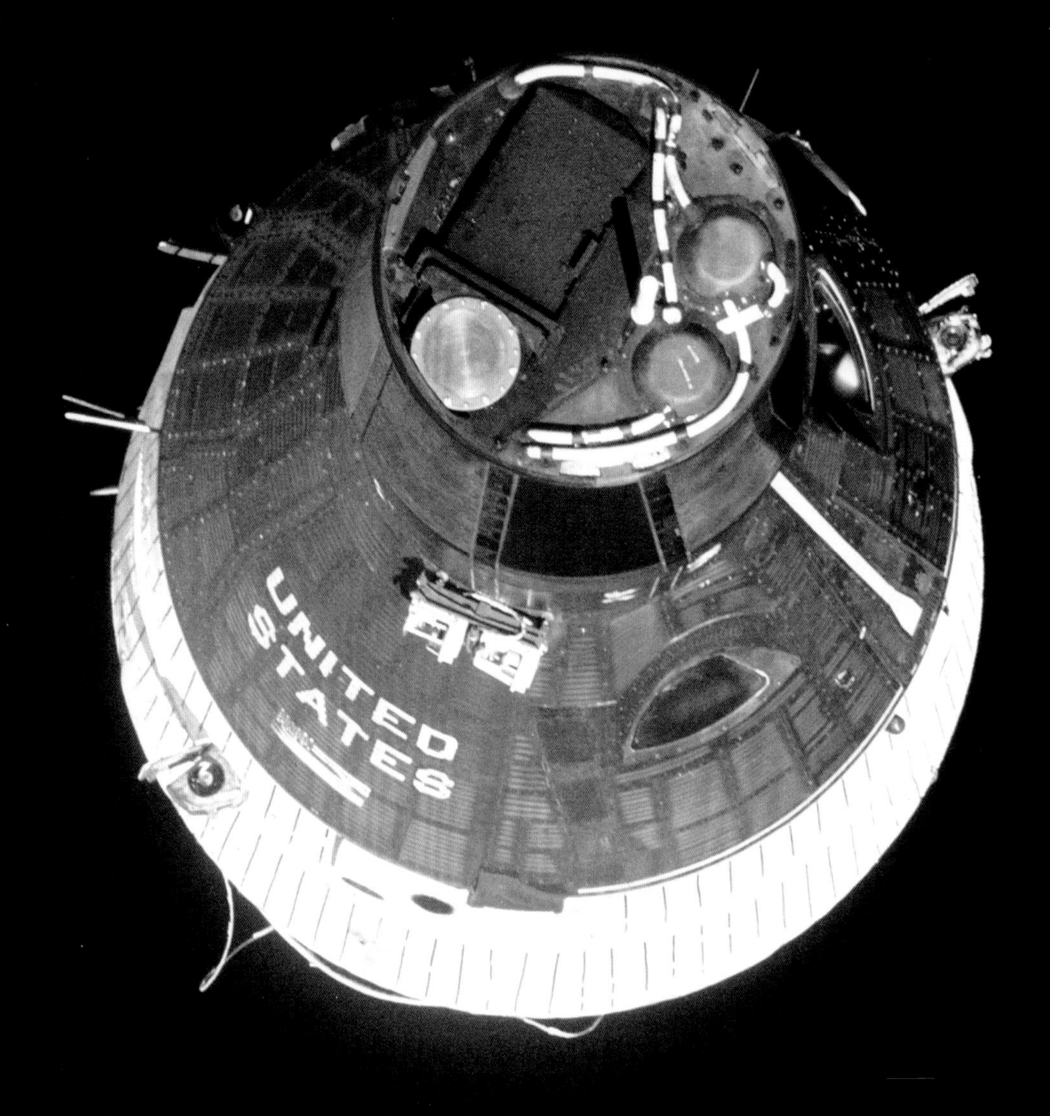

Gemini 6 was to have taken off on 25 October 1965 to attempt a rendezvous with an Agena stage, but the loss of the latter cancelled the mission. At the last minute, NASA decided to send up two manned Gemini craft for a space rendezvous. Gemini 7 lifted off on 4 December. Gemini 6 was to have gone up on the 12th, but the rocket's engines cut out a second after ignition. The flight was then postponed until the 15th. In the meantime, Lovell and Borman, cooped up in their orbiting capsule, were getting thoroughly bored! Eventually, on 15 December, the Titan II rocket took off and carried Stafford and Schirra into space…

Rendezvous in space

While other Gemini missions were intended to practise rendezvous in Earth orbit, the first rendezvous between two Gemini capsules was actually the result of chance! Gemini 6, with Walter Schirra and Thomas Stafford, was supposed to dock with an Agena, but the Titan rocket and its Agena were accidentally destroyed in October 1965 just before the launch of the Gemini. NASA then decided to postpone the flight, renaming it Gemini 6A, and to try instead to dock with the next capsule, Gemini 7. The latter therefore took off before Gemini 6A, on 4 December 1965, with Frank Borman and James Lovell on board. Gemini 7 was the most productive of the series of missions – no fewer than 20 experiments were completed – and it was also NASA's longest before Skylab, its crew being in space for nearly two weeks.

On 15 December, Gemini 6A lifted off in turn, though not without a few scares. Lift-off was initially due to take place on the 12th. On the appointed day, the countdown was completed as normal and the Titan rocket's main engines ignited; but the on-board computer detected a problem and shut down

The two craft manoeuvred together for several minutes. Gemini 6 was sent up to rendezvous with her sister capsule Gemini 7. The detonator wires left trailing after the separation can be seen here, and caused Schirra to call out: 'Hey, Lovell, your capsule looks like a badly tied parcel.' Gemini 6 was the last vehicle in the series to use batteries rather than fuel cells to provide on-board electricity. The two missions were a double success for NASA: Gemini 7 broke the endurance record and the Americans achieved the first real rendezvous, bringing the two craft within a few inches of each other.

The Gemini 7 mission was quite tedious, as the crew complained. There was not very much to do during this record flight, other than take photographs. Here the capsule is flying over the cordillera of the Andes, with the mountains forming a significant barrier to the clouds, readily visible in the picture.

the firing. The usual procedure would have been for the crew to eject, in case the rocket should crash on its launch pad. Schirra, in his capacity as commander, decided not to eject, being sure that the rocket had not moved off the pad and being disinclined to attempt the very risky procedure. The engineers managed to find the fault and Gemini 6A finally lifted off 72 hours later.

In a lower orbit than Gemini 7, Gemini 6A was quickly able to approach it. Radar contact was made three hours and 15 minutes after lift-off, when Gemini 7 was 270 miles ahead. Eventually the two capsules came within three feet of each other and the two crews talked over the radio. During the approach manoeuvre, Gemini 6A used only a tiny amount of fuel, proving that all that was needed for a successful rendezvous was the right trajectory and sufficient speed. Gemini 6A returned to Earth after one day and one hour, its splashdown being the first to be witnessed live on television. Borman and Lovell stayed up for another three days; their timetable was less busy and they were able to spend a lot of time reading. A guidance-rocket fault was flagged up, but did not prevent the mission from continuing. The exit from orbit and splashdown went perfectly.

This is what Armstrong and Scott could see. The Agena is floating in space, waiting for Gemini 8 to close in from the right. The Agena Target Vehicle was approached, but the operation did not last long as both craft began to rotate rapidly. Armstrong took the decision to back off from the ATV but the Gemini capsule continued to rotate on its own. The mission was aborted and a hasty return to Earth had to be considered. Armstrong and Scott were criticised, but Gene Kranz, flight director until Apollo 17, ended the argument by saying: 'The crew reacted exactly as in training and they reacted wrongly because we trained them wrongly.'

Neil Armstrong made his career in the Navy as a bomber pilot in the Korean War and was hit during a low-altitude attack in August 1951, but bailed out over Allied territory. He later became a test pilot at Edwards, and flew the X-1B, the X-20 Dyna Soar and the X-15, making some flights into space. He was the leader of the astronaut group and he was asked if he would like to join the new Gemini team in September 1962. Accepting, he commanded the Gemini 8 and then Apollo 11 missions, becoming on the latter the first man to set foot on the moon. David Scott was part of the third group of NASA astronauts. He flew with Gemini 8 and Apollo 9 as CMP (Command Module Pilot). He commanded the Apollo 15 mission, which took place in July-August 1971. This mission was marred by the 'stamp scandal', when stamps were taken up and cancelled during first day's flight and later sold to a German collector without NASA's knowledge. It was hardly a serious matter, as none of the astronauts made a penny out of it – the money was given to a charity – but NASA decided to make an example and Scott did not fly again.

Docking troubles...

In the late afternoon of 16 March 1966, Cape Canaveral's Launch Pad 19 trembled and was enveloped in white clouds as a Titan rocket, carrying Neil Armstrong and David Scott in Gemini 8, wrenched itself free of the Earth's pull. The two men were to attempt a docking with an ATV (Agena Target Vehicle) and carry out a lengthy EVA. The first part of the mission went as planned. The Agena stage – launched by a Titan more than an hour and a half before Gemini 8 – was located on the capsule's radar, and at about 130ft distance the crew took time to check it from every angle to make sure that it was not damaged. At a rate of just 3in a second, Gemini 8 closed in and Armstrong locked its nose into the ATV's docking ring. It was then that the two craft began a roll that steadily increased. Armstrong undocked the capsule, but the problem worsened, the craft making a full turn once every second. It transpired that one of the thruster engines had failed, and the other engines had to be used to counter this, but under the circumstances it was decided to curtail the mission and, less than ten hours after lift-off, the Gemini 8 capsule was forced to return to Earth carrying two rather upset and nauseous astronauts! The McDonnell engineers never really got to the root of the problem...

Gemini 9A's flight was not a great success either. Thomas Stafford and Eugene Cernan – who had replaced Elliott See and Charles Basset at short notice after they were lost in a plane crash – were down on their luck. Firstly, the Agena ATV docking stage blew up in flight on the 17 May 1966. This time there was an alternative in the shape of the ATDA (Augmented Target Docking Adaptor), a similar vehicle to Agena in its front section, but without an engine and fuel tanks. Launched on 1 June by an Atlas rocket, the ATDA had to wait until the 3 June to be joined by Stafford and Cernan, whose 1 June countdown had been halted. As the two men got to within about 50 miles of the ATDA, they were pleased to see its lights flashing.

Gemini 8 came home early and it was NASA's first emergency return. Scott was to make an EVA, but with the capsule going into a spin everything was aborted. The craft then made an immediate re-entry to the atmosphere, as the two men were finding it hard to get their bearings and read the instrument panel. The retro-rockets were fired over equatorial Africa and Gemini 8 re-entered the upper layers of the atmosphere. Failure had resulted from being unable to come up with a solution to a problem. It would serve as a lesson for Apollo 13...

The Agena crocodile

Unfortunately, it wasn't long before they noticed that the cap over the ATDA's nose had not been properly jettisoned. Half open, it made the craft look like 'an angry alligator' according to Stafford. The explosive release-bolts on the cap had worked correctly, but someone had forgotten to remove the straps holding the two sections of the cap together! Abandoning the docking, the two men concentrated on the spacewalk. Cernan had difficulty getting out of the capsule despite the presence of handholds on the outside that had been added following Ed White's comments after the previous mission. The astronaut was also charged with trying out the AMU (Astronaut Manoeuvring Unit), a back-mounted propulsion system designed to allow the astronaut to move around independently. This 'backpack' was stowed in the Adaptor Module compartment, but it was difficult to get at, and Cernan couldn't manage to put it on. He was sweating so much that his helmet was misting up and his movements were hampered by the umbilical cord linking him to the capsule. Eventually he decided to give up and return to his seat in the capsule. After this 128-minute trip, Gemini 9A returned to Earth with no further problems.

Gemini 10 had better luck. Piloted by John W. Young and Michael Collins, the capsule docked successfully with an Agena on 18 July 1966. Because of the latter's engine and fuel tanks, Gemini 10 was able to continue on a much higher orbit (474 miles, a record), before descending again to find the Agena stage abandoned after the Gemini 8 mission. Collins went outside, remaining close to the airlock and taking a few photographs; then, after repressurising the capsule, the two men took a nap. The next day they undocked from their Agena booster and approached Gemini 8's. With the capsule less than ten feet away, Collins made a second EVA. He tried to get hold of the craft, but there was nothing to cling on to. Thanks to his gas-gun, he was able to move around easily and recovered a micro-

The crew for the ninth Gemini mission consisted of Stafford and Cernan. Eugene Cernan would go into space three times: on Gemini 9, Apollo 10 and Apollo 17. He would do an EVA with Gemini and was the last man on the moon in December 1972. Of Czech origin, he was a Navy test pilot.

'It looks like an angry alligator!' exclaimed Stafford on finding the ATDA (Augmented Target Docking Adaptor). The explosive bolts on Agena's nosecone had fired, but it was still held by a piece of strapping. This had been detected from the ground, but it was assumed that a simple EVA would remedy the problem. However, nothing could be done and Gemini 9, just like Gemini 8, was unable to dock. Nonetheless, the capsule approached and moved away from the target vehicle several times as it slowly spun. Within 24 hours the crew had carried out three close-encounter rendezvous.

This curious stroboscopic photograph shows the launch of Gemini 10 and the way the launch tower (or Erector) is lowered in the seconds before lift-off, while the 'umbilical' tower remains fixed. This launch pad is Launch Complex 19 (LC 19) at Cape Canaveral. Constructed uniquely for the Gemini launches, it was abandoned once the missions were over.

meteorite collector. Back on board, he and Young began a series of experiments before setting foot back on Earth on 21 July.

Gemini 11, crewed by Pete Conrad and Richard Gordon, was notable for its rapid docking with the Agena stage, only a few minutes after lift-off, and also for the highest orbit – 853 miles – ever achieved by man, if the Apollo moon missions are excepted. Gordon made two sorties into space, but came up against the same problems as his predecessors. By contrast, the approach manoeuvres to the Agena were carried out several times without any problems and, for the first time, the return was carried out entirely on automatic.

The final Gemini mission, Gemini 12, took off on 11 November 1966 with James Lovell and Edwin 'Buzz' Aldrin aboard. This time the EVAs had been better planned: swimming-pool simulations had been adopted in training, which were much more effective than those carried out in the 'Vomit Comet' planes where just a few seconds of weightlessness were achieved by flying in a parabola. Aldrin made a full space-walk of a little over two hours, followed by two others where he remained near the hatch. Thanks to the new underwater simulation training method (which was to become the norm) the EVAs proved much easier to manage.

With Gemini, NASA had completed the preliminary phase of its mission. Its next objective was the moon.

A view of Agena a few cables' lengths away from Gemini 12. The docking was carried out manually after a total failure of the radar. The Agena was then supposed to take the capsule into a higher orbit, but because ground control had detected a problem with the rocket's engine it was decided not to restart it.

Aldrin made three EVAs during the Gemini 12 mission. The Gemini astronauts' padded helmet was not very well suited to EVAs, with its limited field of view. It was not until the Apollo EVAs that the well-known 'goldfish bowl' helmet with its all-round vision came into general use.

Cernan in the capsule before his EVA. He had to go to the rear of the capsule, where the AMU was stored. This was a kind of chair fitted with thrusters, intended to allow an astronaut to move around freely, but its use proved difficult. The lack of proper handrails (the engineers had, however added some since White's EVA) and its reduced rigidity were handicaps: when Cernan lifted his leg, his whole body tipped backwards. The 'snake' or umbilical cord that linked the astronaut to the capsule also got in the way. Furthermore, when the sun disappeared behind the Earth Cernan's visor began to fog. Cernan eventually gave up, with mission control's consent. When he rejoined Stafford, the latter could barely see his colleague's face there was so much mist on the visor!

Like the Mercury capsule, Gemini was braked by a simple parachute. To lessen the shock, it came down almost horizontally, with the two astronauts lying on their seats. Frogmen then had to be sent in to correctly fix the inflatable U-shaped raft that surrounded the capsule. The hatches were then opened, but the crew had to wait to be lifted aboard a ship. This time it was the carrier *Wasp*, an old hand in NASA's recovery teams, that picked up Cernan and Stafford.

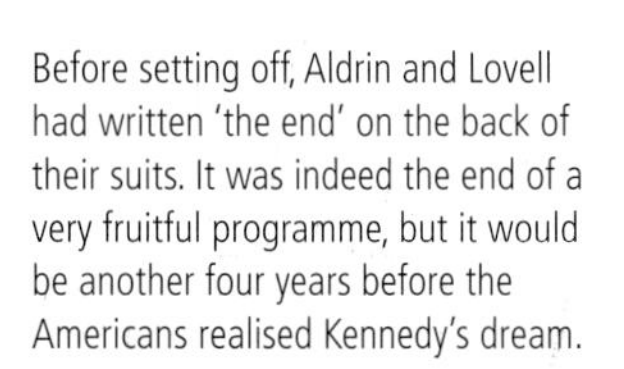

Before setting off, Aldrin and Lovell had written 'the end' on the back of their suits. It was indeed the end of a very fruitful programme, but it would be another four years before the Americans realised Kennedy's dream.

The ingenious Agena

The Agena system has been used equally well as an upper stage on Thor (military), Titan and Atlas rockets to put a whole series of civil and military satellites into orbit. This flexibility provides it with the capacity to adapt to almost any kind of launch, particularly as it became highly reliable once fully developed. Of military origin (its Air Force designation was RM-81), this rocket was intended to send up military reconnaissance satellites, but also to have a certain range once in orbit. This 'two-in-one' concept was novel, but the programme was eventually abandoned. Agena subsequently found its niche in the need to increase the payloads of contemporary launchers: most of the low-cost launchers could be fitted with this 'turbo', thereby allowing them to put into orbit loads that they could not have managed on their own. Obviously, this simple solution obviated the need for a more powerful and more expensive rocket.

Constructed by Lockheed, the Agena A, used with the Thor and Titan rockets, became a real 'space tractor' in its B version, whose Bell XLR81 engine – one of the rare engines to be made from aluminium – was more powerful and could be restarted in orbit. Finally, the Agena D was a standardised version of the Agena B, capable of adapting to a multitude of satellites and spacecraft without modification. This version, known as the ATV (Agena Target Vehicle) during the Gemini programme, was modified by the addition of a docking system designed by McDonnell to allow docking manoeuvres to be tested in space. Furthermore, this version could be radio controlled by the Gemini astronauts or from a ground station using two small auxiliary engines. Used in this way, Agena was launched on top of an Atlas rocket, with Gemini following (at least 90 minutes later). The ATV was then used as a target vehicle for a space rendezvous to test docking procedures. When the docking had been completed, the vehicle could then move into a higher orbit using its XLR81 engine. There was, however, a difference between the Agena ATDA (Augmented Target Docking Adaptor) tested during the Gemini 9 mission, and the ATV used in Geminis 8 and 10, the ATDA not having an engine that could be restarted as this was not thought necessary. In total, taking into account all versions, 365 Agenas were launched into space. Not a bad score!

Technical details

Name: Agena D

Type: upper stage

Country: United States

Manufacturer: Lockheed

First launch: 28 June 1962

Last launch: 12 February 1987

Power on lift-off: 71.17kN

Engine:
1 Bell XLR81-BA-9 rocket engine running on a liquid-nitric-acid/UDMH mixture for 265 seconds

Length: 24.8ft

Payload: 0lb

Total weight: 70.5 tons

Launch of an Atlas-Agena rocket.

Chronology of the Gemini and Voskhod programmes

Mission	Launch date	Comments
Voskhod 1	12 October 1964	First orbital flight with three people.
Voskhod 2	18 March 1965	First spacewalk, by Leonov.
Gemini 3	23 March 1965	First orbital American flight with two astronauts.
Gemini 4	3 June 1965	First American EVA, with White.
Gemini 5	21 August 1965	First long-endurance flight.
Gemini 7	4 December 1965	Record flight length and rendezvous with Gemini 6A.
Gemini 6A	15 December 1965	Rendezvous with Gemini 7.
Gemini 8	16 March 1966	Docking with Agena stage; partial failure, cancellation of Scott's EVA, emergency return to Earth.
Gemini 9	3 June 1966	Attempted docking with Agena abandoned; Cernan's EVA, MMU not used.
Gemini 10	18 July 1966	Two rendezvous with two Agenas, one docking and engine firing; two successful EVAs.
Gemini 11	12 September 1966	Docking with Agena; firing of the craft's engines to take it to highest altitude ever achieved (853 miles); first fully automated re-entry.
Gemini 12	11 November 1966	Long EVAs, rendezvous and docking with Agena, fully automated re-entry.

Crew	Duration of flight
Komarov-Feoktistov-Yegerov	1 day
Belyayev-Leonov	1 day and 2 hours
Grissom-Young	4 hours and 52 minutes
McDivitt-White	4 days and 2 hours
Cooper-Conrad	7 days and 22 hours
Borman-Lovell	14 days
Schirra-Stafford	1 day 1 hour and 51 minutes
Armstrong-Scott	10 hours and 40 minutes
Stafford-Cernan	3 days and 20 minutes
Young-Collins	2 days 22 hours and 46 minutes
Conrad-Gordon	2 days and 23 hours
Lovell-Aldrin	3 days and 22 hours

From Earth to the moon p. 69

The moon within reach... p. 86

Achievement becomes routine p. 103

CHAPTER 3

From Earth to the moon

Third and final stage on the great leap to the moon, the Soyuz programme and its American equivalent Apollo were complex, ambitious and...exceptionally expensive.

The two superpowers had to develop much more powerful launchers than any that existed at the time. They could not draw on their military arsenals, the converted intercontinental missiles not developing sufficient thrust to propel a craft – indeed, a collection of craft – into a sufficiently high Earth orbit to ensure a successful trip to the moon. Consequently huge sums were eaten up in the design, development and construction of new rockets. The Soviets developed the N1, a 3,000-ton monster that met with almost total failure, while on the American side it was the Saturn rocket that would be successfully entrusted with taking its astronauts to the moon. The Saturn V version remains to this day the most powerful operational rocket ever constructed.

Soyuz eyes up the moon

The Soviets sent numerous Luna probes to the moon between 1958 and 1976. Some of these missions were failures, but Luna 3 managed, for the first time, to take pictures of the dark side of the moon in 1959. Then in 1966 Luna 9 made a soft landing on the moon's surface. The programme accelerated in response to Kennedy's 1961 challenge, but Voskhod lacked the potential for this kind of mission and a new vehicle had to be developed. From 1963 Korolev opted for a

On 25 May 1961, after being in office only since January, John Fitzgerald Kennedy delivered the memorable speech that would give rise to the Apollo programme. He asked Congress to release emergency funding. In September 1962, in the stadium at Rice University, he repeated: 'We have chosen to go to the moon within this decade and to do other things [in space], not because these things are easy, but because they are hard...'

Above: Training for astronauts likely to go to the moon on the Apollo missions was noticeably more exhaustive than for the other programmes. They had to know like the backs of their hands not only the flight procedures for the two craft (the command module and the lunar module), but also how to handle the various tools for collecting samples. Geology was an important aspect of the Apollo programme and here Shepard can be seen with the MET rover, busy collecting rocks. According to the other astronauts, geology was not Shepard's favourite subject!

Above centre: Collins in Apollo's command module simulator. The position of every switch, dial and button had to be known, including in dark or smoky conditions, in a craft liable to be tossing its occupants around in every direction. Nothing was left to chance. These incessant rehearsals were both indispensable and effective. All the Apollo missions that went into space brought their crews back safely.

Above right: Shepard in the middle of a training session in a Vomit Jet, a KC-135, which briefly reproduced the conditions of weightlessness.

Armstrong and Aldrin training on a simulation of the moon's surface, with a life-sized model of the lunar module in the background. Every movement is closely observed by a team of specialists who check the slightest detail: if a tool was the wrong size the collection of samples could become a problem.

Soyuz-type set-up for the lunar programme, which was named 'Zond'. Soyuz ('Union') consisted of a piloted craft (Soyuz A) with living space, the return capsule and the propulsion module; an engine module with empty tanks (Soyuz B); and a unit containing three fuel tanks (Soyuz V), the last a real space tanker. These three units were to be launched separately. Soyuz B would be the first to go up, followed by Soyuz V. They were to dock automatically in space and, after transferring its fuel, Soyuz V would be jettisoned. The piloted Soyuz A would then lift off and join up with Soyuz B in Earth orbit, before heading for the moon. In practice, this plan was not pursued. Various Soyuz tests ended in disaster, and the original designated moon rocket, the N1, never became operational. The Soviets thus abandoned any idea of going to the moon, but the Soyuz spacecraft itself survived the moon programme to become the most used vehicle for getting into space. In modernised form, Soyuz is still in service today.

The Soyuz spacecraft was rather different from the early Vostoks. The first generation of craft (from 1967 to 1971) consisted of three distinct units with, from front to rear, the spherical orbital module housing two or three cosmonauts; the smaller re-entry module; and the service module with its distinctive solar panels. Soyuz proved to be more comfortable than its American competitor Apollo. It was also notably lighter, as only the small re-entry module was burdened by the weight of a heat shield, the orbital and service modules not returning to Earth. Furthermore, each of the two modules could be isolated and depressurised, thus serving as an airlock for any EVAs. While the re-entry module was used

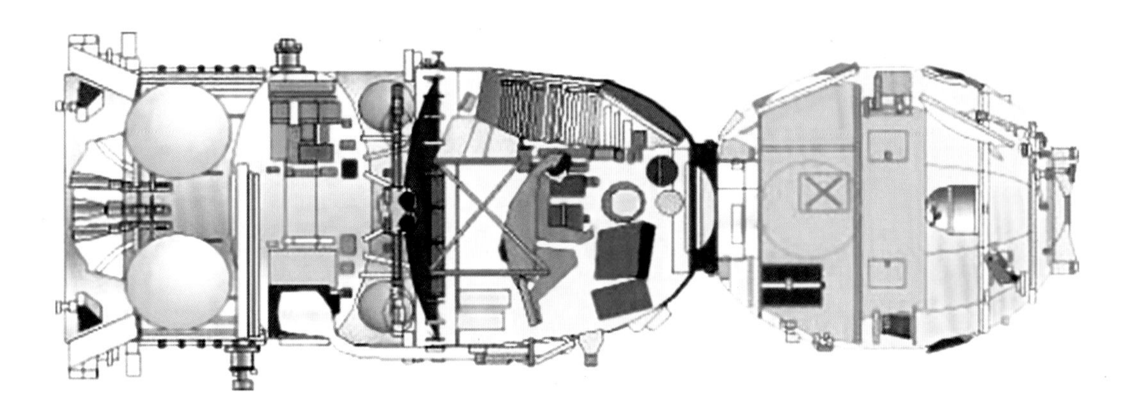

The Soyuz craft as it appeared in orbit. Right at the front is the orbital module. Behind that, the descent or re-entry module (used only when ascending and during the return to Earth) and, finally, the service module where all the instrumentation and equipment was located.

for eating-breaks, the orbital module was favoured for accommodation. Unlike Vostok and Voskhod, the re-entry module was not spherical. A spherical craft would have been capable of making only a ballistic re-entry, creating no lift. Any cosmonaut inside suffered a high g-force level and landing was rough. With its 'car headlight' shape the Soyuz descent module generated more lift and, apart from the main parachute slowing it down, retro-rockets would fire for a second when it was a metre from the ground. Finally, a number of procedures were to be automated, including the approach to other spacecraft.

On paper, Soyuz was a very appealing spacecraft. Designed by Korolev's research office (Korolev himself died in 1966), it was pressurised with a mixture of oxygen and nitrogen, whereas the Apollo capsules used pure oxygen. The first Soyuz model was the 7K-OK. It was designed for three men and could dock only with another Soyuz of the same type, although transfer between them had to be made via an EVA. Designed for Earth orbit, the 7K-OK encountered a number of problems during its unmanned test flights starting in November 1966. But, just as with the Americans, time was pressing and they needed to make a manned test. Korolev's successor, Vasily Pavlovich Mishin, was asked by Brezhnev himself to speed up the programme, since, with the American Gemini programme in full swing and Apollo taking shape, the Soviets had fallen somewhat behind in the race. The first manned flight was planned for April 1967.

The first manned Soyuz flight ended fatally for Komarov. It was not actually the craft that failed, but the parachutes. With no means of escape, Komarov was a condemned man, and crashed into the ground a few seconds after re-entry. He was from the initial group of cosmonauts and had already flown in Voskhod 1.

While not the inventor of the LOR (Lunar Orbit Rendezvous) concept, the engineer John C. Houbolt, who worked at the Langley research centre, managed to get the idea accepted by NASA's directors. The concept brought success to the Apollo programme by saving time. The three-part Apollo was based on the command module linked to a service module (like Gemini) and the freestanding lunar module. The three systems were not especially heavy, so that only one launcher, thus less fuel, was needed. Houbolt's idea was challenged by two other concepts. Firstly, there was the Nova super-rocket, which would take off from Earth, fly to the moon, land and then return, but this direct flight approach never got beyond the drawing-board stage. The second idea, the EOR (Earth Orbit Rendezvous), came from von Braun and the Marshall Flight Center. Two rockets would take two elements of a station up into low orbit preparatory to a rendezvous with a pre-assembled lunar module. In orbit, and with docking achieved, the crew would transfer to the lunar module, which would transport them to the moon, land, then return to the station. Much to von Braun's displeasure, NASA opted for Houbolt's solution.

Sadly, Soyuz 1 was to be the first Russian space disaster, and the second (following the Apollo 1 accident described further on) in the history of the conquest of space. Soyuz 1 and cosmonaut Komarov were to lift off first and then be joined by Soyuz 2 with three more men aboard. The two craft were to dock and two members of the Soyuz 2 crew were to join Komarov. But on 23 April 1967 everything went wrong.

As soon as Soyuz 1 was in orbit one of its solar panels refused to open. Despite various attempts, nothing would make it work. As there was insufficient electricity, the decision had to be made to either send up Soyuz 2 immediately so that the cosmonauts could get outside and release the panel, or to bring Komarov straight back. Unfortunately the first option had to be abandoned because of terrible weather over the Baikonur launch site. So, after a day in space, Soyuz 1 began its re-entry, although with some difficulty as the retro-thrusters were not working properly. But the worst was still to come. The small drogue parachute opened first and slowed the spacecraft a little. Normally the drogue would be jettisoned at this point, to avoid its getting entangled with the main parachute, but because of the very narrow compartment where it was stowed the main parachute failed to unfold. Komarov released the emergency parachute, but it got twisted up with the drogue chute that was still attached. Soyuz 1 crashed into the ground and Komarov was killed instantly.

The Soyuz programme had difficulty getting going again after this accident, especially as the engineers discovered that the Soyuz 2 spacecraft, which didn't finally take off until 1967, suffered from exactly the same defects. As the result of an enquiry around 200 defective parts had to be replaced and the Soviets lost nearly 18 months. It might as well have been an eternity. They were now running far behind schedule, as Soyuz 2 did not get going again until October 1968, by which time NASA was well into its Apollo programme.

Uncle Sam's third programme

Looked at prosaically, the Apollo programme consisted of a series of manned flights by the United States, with the aim of putting a team on the moon and getting them safely back again. This aim was achieved in July 1969, five months before the date set by President Kennedy. Yet, in the collective imagination, 21 July 1969 represents much more than just a simple exploit. The ensuing flights, even though more complex from a technical point of view, would never have the same impact. Going to the moon was not just Kennedy's idea brought to fruition by NASA. The programme had actually been conceived during the presidency of Eisenhower (who won two terms in office, from 1953 to 1961, and initiated the Mercury project). Apollo was supposed to have followed Mercury, but to ensure a solid preparation for the mission the Gemini programme was inserted between them. For Apollo, various avenues were explored. The first was to send a spacecraft directly from Earth to the moon, land, and then return intact. For this, a very powerful launcher would be required – the engineers came up with the Nova rocket, which never saw the light of day.

The second route considered was the EOR (Earth Orbit Rendezvous), involving the more-or-less simultaneous launch of two Saturn rockets, one carrying the spacecraft, the other its fuel – a similar principle to that used by Soyuz – the spacecraft then docking with its fuel tanker in orbit before heading off to the moon. Here again there was only one spacecraft involved: it was to fly to the moon, touch down and return.

A third option was the LOR (Lunar Orbit Rendezvous). A single Saturn rocket would put two vehicles into Earth orbit: a CSM (Command and Service Module) and an LM (Lunar Module). The two craft would rendezvous in Earth orbit, but only the CSM would be manned until they arrived in lunar orbit. At this point, two men would go aboard the LM, land on the moon and walk on its surface before getting back on board. The LM, consisting of two parts, a descent module and a return stage, would then split into two, as the descent module was no longer of any use and would be left on the moon. The upper part of the LM would then take off, rejoining the CSM. Although the LOR may seem to be more complicated, involving many more manoeuvres, it had an enormous advantage: it would save precious weight, given that only a small part of the whole unit (the LM) would land on the moon and only the CSM would be required for the return. Furthermore, only one Saturn rocket was needed. In short, both time and weight would be saved by adopting the LOR option – the Saturn rocket being already available, unlike the Nova – and so too was money, a sledgehammer argument bearing in mind the massive cost of the programme.

The Apollo programme was to consist of 11 manned flights (from Apollo 7 to Apollo 17) and three test flights. There were officially no Apollo 2 or Apollo 3 flights, while the Apollo 1 designation was applied retrospectively to the craft destroyed in the 1967 accident, and flights Apollo 4 to 6 were all unmanned. Other flights had been planned, but NASA's budget was revised substantially downwards, in part to work on the Space Shuttle (whose main object was to reduce costs – but also to put Skylab into orbit), itself an extension of Apollo.

The famous lunar module (LM), also known as the LEM (Lunar Exploration Module) gave NASA and especially its designer, Grumman, a lot of headaches. Its development was long, laborious and costly, delaying the programme in the process. To test the machine, several modules were constructed by the Bell company: two LLRVs (Lunar Landing Research Vehicles) and three LMTVs (Lunar Module Training Vehicles). Able to take-off and land vertically with great precision, they were used to simulate a lunar landing, but proved to be unstable. Three of them were wrecked in accidents and Neil Armstrong himself had to eject during one training session.

The Saturn-Apollo duo

The Apollo programme really got under way on 27 October 1961, a few weeks before John Glenn's flight into space. On that day, the Saturn I rocket made its maiden flight (mission SA-1). To understand the development of this launcher, it must be understood that at this time NASA was still in its infancy (it had started in 1958) and most of its work on rockets was still at the planning stage. NASA had on its drawing board a super-rocket named Nova, but none of its engines had yet been tested. Von Braun's research department at the Redstone arsenal was particularly active and had made considerable contributions to the development of ballistic missiles for the American Department of Defense. In 1957, von Braun had presented the draft sketch of a launcher capable of putting a payload of ten to twenty tons into Earth orbit (or five to six tons elsewhere in the solar system). This outline, named 'Juno V', was finally approved by ARPA (Advanced Research Project Agency), an agency working for the Department of Defense and charged with seeking new technologies that might be of use to the Army. It would become noted for the development of the Arpanet network that formed the basis for computer networking.

But the Army did not really need a super-launcher and the project only got going properly when President Kennedy gave the green light for sending a man to the moon. When, therefore, NASA got von Braun and his team together it found itself with two rockets: Nova and Juno V, renamed Saturn. The latter was eventually chosen because it was easier to construct, but Nova was kept on in case Saturn should fail. These launchers, designed along with all their systems at the Marshall Space Center, actually included three families: Saturn A (A-1 and A-2), Saturn B (B-1) and Saturn C (C-1, C-2, C-3, C-4 and C-5). These families differed in the engines used for the various stages that were put together according to requirements. Saturn A-1 was actually just a cylindrical assembly of eight Redstone rockets around a tank from a Jupiter. These eight H-1 engines were improved by the firm Rocketdyne (a division of North American) to give each one a thrust of 95 tons. On Saturn IB (or B-1), Rocketdyne modified the F-1s to get even more thrust (103 tons) so as to be capable of carrying even more powerful upper stages. Finally, Saturn V (C series) was designed around new F-1 engines with a huge thrust of 677 tons each. Doing a simple sum, it can be seen that Saturn IB generated a first-stage thrust of 827 tons on lift-off, while Saturn V achieved the record figure of 3,385 tons of thrust! In addition, the final J-2-engined stage could be reignited to send the CSM on its correct lunar trajectory.

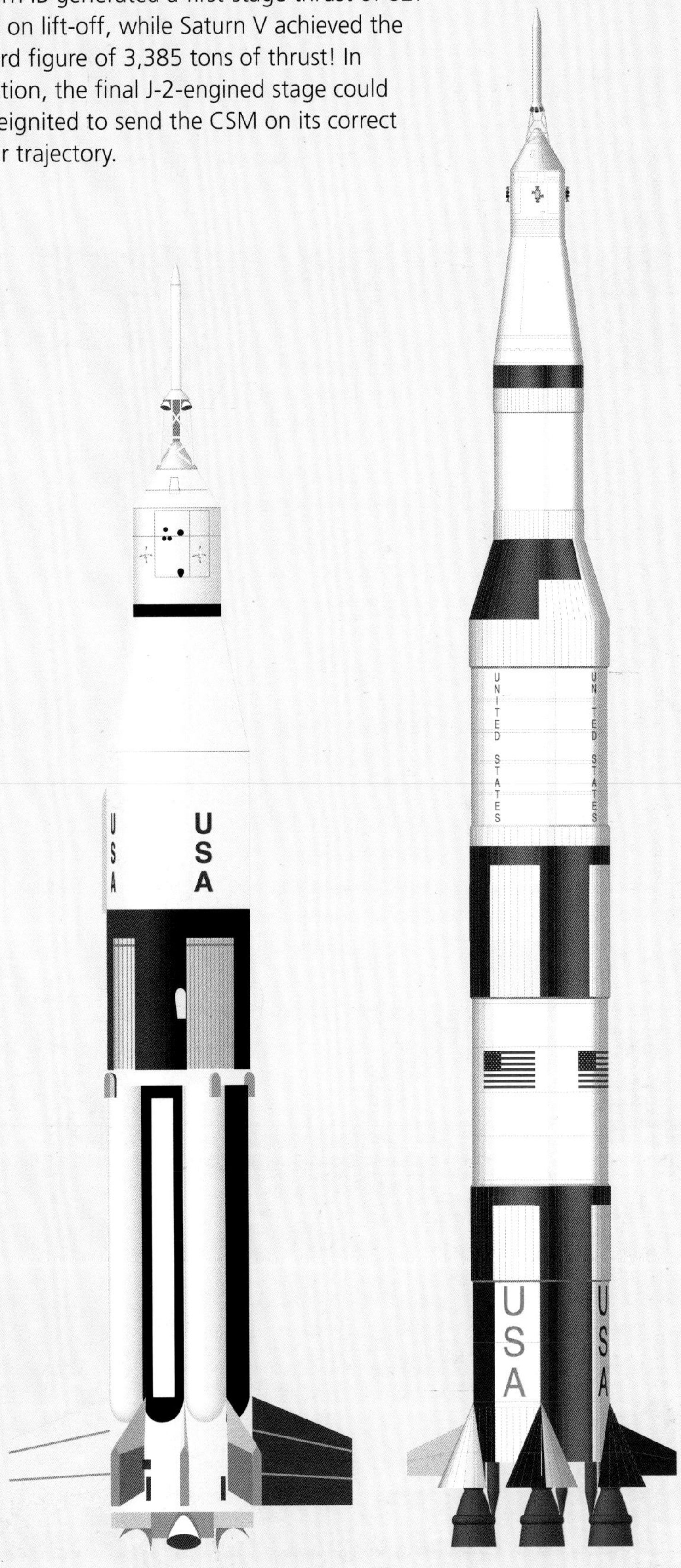

The Saturn launcher family

	1st stage	2nd stage
Saturn A-1	8 H-1 engines running on a mixture of liquid oxygen and kerosene	2 LR-87-3 engines (from the Titan rocket) running on a liquid oxygen/kerosene mixture
Saturn A-2	8 H-1B engines running on a liquid oxygen/kerosene mixture	4 S-3 engines (from the Jupiter rocket) running on a liquid oxygen/kerosene mixture
Saturn B-1	8 H-1B engines running on a liquid oxygen/kerosene mixture	4 S-3 engines (from the Jupiter rocket) running on a liquid oxygen/kerosene mixture
Saturn C-2	8 H-1 engines running on a liquid oxygen/kerosene mixture	4 J-2 engines (Saturn S-II) running on a liquid oxygen/liquid hydrogen mixture
Saturn C-3	2 F-1 engines running on a liquid oxygen/kerosene mixture	4 J-2 engines (Saturn S-2-C3) running on a liquid oxygen/kerosene mixture
Saturn C-4	4 F-1 engines running on a liquid oxygen/kerosene mixture	4 J-2 engines (Saturn S-2-4) running on a liquid oxygen/kerosene mixture
Saturn C-5 (Saturn V)	5 F-1 engines running on a mixture of liquid oxygen/kerosene	5 J-2 engines (Saturn S-2 C-5A) running on a liquid oxygen/liquid hydrogen mixture

The first stages of the Saturn rockets were built on a production line. Here, the first stages of Saturn IB are being prepared at Boeing's Michoud, Louisiana, plant. This complex currently makes the Shuttle's external tank. The nozzles of the eight H-1 engines, typical of this series of launchers, can readily be seen.

In the second photograph, taken in 1968, the same plant is assembling the S-IC stages of the Saturn V launcher. The F-1 engines are significantly bigger and their number has been reduced to five. However, they deliver four and a half times as much power. The four engines around the outside are movable and control the rocket's attitude.

The programme had envisaged three other lunar missions after Apollo 17, but NASA's drastic budget restrictions, as well as the development of the Shuttle, brought a premature end to these missions. In fact, the Saturn launcher was so expensive it was decided not to build a new series. Consequently the available rockets were used for Skylab…or ended their careers as museum exhibits.

While the Saturn rocket is inseparable from the Apollo programme, there was also the 'Little Joe' launcher, which was used to test the capsule's safety tower. However, it was the Saturn 1 and 1B launchers that were used to carry out the programme's first Earth-orbit tests. Saturn V was used to put the craft into lunar orbit. The Apollo programme therefore truly got under way in October 1961 with the first Saturn launch (mission SA-1). Other proving launches were to follow (nine in all) – the A-103, A-104 and A-105 programmes were even used to put Pegasus satellites into orbit. In the summer of 1963 the module's LES

3rd stage	*4th stage*
2 RL-10A-1 engines (from the Centaur C rocket) running on a liquid oxygen/liquid hydrogen mixture	
2 RL-10A-1 engines (from the Centaur C rocket) running on a liquid oxygen/liquid hydrogen mixture	
6 RL-10 engines (Saturn V) running on a liquid oxygen/liquid hydrogen mixture	2 RL-10A-1 engines (from Centaur C) running on a liquid oxygen/liquid hydrogen mixture
6 RL-10 engines (Saturn IV) running on a liquid oxygen/liquid hydrogen mixture	2 RL-10A-1 engines (from Centaur C) running on a liquid oxygen/liquid hydrogen mixture
6 RL-10 engines running on a liquid oxygen/liquid hydrogen mixture	
1 J-2 engine (Saturn IVB) running on a liquid oxygen/liquid hydrogen mixture	
1 J-2 engine (Saturn IVB C-5A) running on a liquid oxygen/liquid hydrogen mixture	

(Launch Escape System) tests were carried out. These were to last nearly three years, followed by the 201, 203 and 202 (in the order they took place) AS (Apollo-Saturn) missions. The unmanned flights came next with the Apollo 4, 5 and 6 missions between November 1967 and April 1968. The main aim of all these missions was to test the Saturn 1B launcher, the CSM (Command and Service Module), the LM (Lunar Module) and finally the huge Saturn V launcher.

The first S-IC stage of Saturn V is lifted upright at the Michoud plant. It is 138ft long with a diameter of 33ft. The second and third stages were manufactured in California, the former by North American and the latter by Douglas. The three elements were then transported to Cape Canaveral to be installed in the VAB (Vertical Assembly Building).

To manufacture and assemble the world's biggest rocket, they had to create the world's largest building. The VAB rises 525ft above Merrit Island, north of Cape Kennedy. It is 518ft wide and 716ft long, allowing each element of the launchers to be erected. It was possible to prepare four rockets at the same time. To bring the rocket out, a huge caterpillar-tracked vehicle, the Crawler Transporter, is placed under the platform and raises Saturn's 3,000 tons and the hundreds of tons of the umbilical tower and takes them to the launch pad.

The first photograph shows a general view of the installation. The building visible at the bottom of the VAB is the firing control centre. The rocket on the Crawler is a Saturn 500F, a test launcher that will never fly, but was used to check various procedures before a launch. The second photograph, taken in 1971, shows Apollo 15 leaving the VAB.

Inside the VAB, the Apollo 501 rocket (Apollo 4 mission) is nearly complete. We see it here placed on its platform, with the assembly structure, which will remain inside the building, to the right. The Crawler Transporter will bring in a mobile platform.

UNITED
STATES

Little is left of the interior of Apollo 1's command module. Everything was burnt up in a few seconds. The speed with which the fire broke out prevented the crew from being rescued and they were asphyxiated. Grissom, White and Chaffee were training for a rehearsal of the flight they should have made a few weeks later. The presence of pure oxygen exacerbated the effect of the fire and NASA did consider changing to a mixture of oxygen and nitrogen.

The Apollo 1 tragedy

The Apollo 1 designation was not originally allocated to a mission, it being applied only after the tragedy of 27 January 1967. What became Apollo 1 was actually mission AS 204. It should be pointed out in passing that there were two official designations for the Apollo flights. The Kennedy Space Center (named in honour of the President in 1962) called its missions 'Apollo-Saturn' (AS), whereas the Marshall Space Center, charged with assembling the launcher, called its missions 'SA' (Saturn-Apollo). Mission AS 204 was thus to have been the first manned flight of the Apollo programme.

What was taking place on the day of the tragedy, on launch pad 34, was just a rehearsal, the actual launch not being due until a month later, on 27 February. The CSM in which Virgil 'Gus' Grissom, Edward White and Roger B. Chaffee were sitting was a first-series craft, designed for orbital rather than lunar flights, as it lacked a means of docking with the LM. The three men were going through a checklist when Chaffee was heard to shout 'We've got a fire in the cockpit!' A few seconds later, the engineers heard shouting and then saw, on the TV monitor, White trying to open the airlock...then everything went black. In an atmosphere composed of pure oxygen, the fire instantaneously became an inferno. It took five minutes for the rescue crews to open the airlock and put out the fire.

The subsequent enquiry established that a simple short-circuit from a poorly insulated wire had caused the disaster. North American, the module's manufacturer, suggested using explosive bolts for the airlock release and, above all, abandoning the use of pure oxygen in the cockpit. But NASA refused, as the bolts could fire prematurely – which was what had happened to Grissom during splashdown following his Mercury flight – and pure oxygen was a reliable, simple means of ensuring the crews' life support in space. Eventually, however, all the Apollo cockpits were revised to offer greater safety in the event of a fire on board. Non-flammable materials were used, wiring had improved insulation, oxygen pressure was reduced and the airlock hatch was made outward opening. The still-intact Saturn 1B launcher was disassembled and re-erected on launch-pad 37B, to be used for Apollo 5.

If something went wrong in the first seconds of the flight, the only survival system the crew had was the rescue tower. The Saturn rocket was much more dangerous than Gemini's Titan, as the propergols used were highly explosive (Titan's would burn, but not explode). As with Mercury, Apollo's command module was topped by a tower fitted with four nozzles giving a six-second burst of 70 tons thrust. This amount of time was needed to put the capsule into an escape trajectory. After separating from the tower, the module came down using its parachutes.

The CSM came back down ten hours after the launch. It was picked up by USS *Okinawa*. As can be seen, the exterior layer was well and truly 'cooked'. The module was made from two shells, fitting one inside the other. The outer shell was made from stainless steel with two honeycombed layers. This was covered in an ablative material, which means that it peeled like an onion when subjected to heat. The inner shell, housing the crew, was insulated from the outer part by quartz fibre matting. The total weight was no more than 4.3 tons.

Apollo 4 took off on 9 November 1967. It was the first unmanned flight in its more or less definitive form, although the lunar module was replaced with ballast. The lift-off was attended by much greater vibration and noise than had been expected, even though launch pad number 39 – where Saturn V was taking off for the first time – had been built more than four miles away from the VAB and the launch control centre.

This view of the separation of one of the elements joining Apollo 6's two stages has been seen around the world. There were one or two problems with the first- and second-stage engines on this mission. Engine 2 then engine 3 of the second stage cut out prematurely, but the on-board computer compensated by increasing the power and the duration of burn of the remaining three engines. Finally, the last stage of the S-IVB did not re-ignite once it was in orbit. It was supposed to have allowed the simulation of a trans-lunar insertion. With the CSM (Command and Service Module) unoccupied, it was decided to use its engine to get into a lunar trajectory, albeit in a much higher orbit.

The Apollo 5 mission was used to test the LM's engine by putting it into a low orbit. The LM had not been fitted with its 'spider legs', as these would not be required. Here it is being lowered into its housing before having the Saturn IB's nosecone fitted on top. The launch of this flight took place on 22 January 1968 instead of 5 April 1967 as originally planned. The engines' development consumed much time and during tests one of the portholes cracked under pressurisation. These were soon replaced with simple pieces of aluminium. The flight eventually passed off very well, to NASA's great relief.

Apollo 7's crew (from left to right Don Eisele, Walter Schirra and Walter Cunningham) pose in front of the command module's access hatch. The mission was peppered with clashes between the flight director and the crew, who were, in consequence, dropped from the rest of the programme.

The programme moves on...

NASA was much criticised, especially as the tragedy had happened on the ground, not in space, so it was absolutely essential to achieve a successful flight in order to maintain credibility and be sure of continuing public support. Apollo 4 was the Saturn V's first flight, the rocket being fired from launch pad 39, which had been specially built for it. This unmanned flight was most important, since the costly and delayed Saturn V was to be tested in its final configuration, with a new second-series CSM, and ballast weight in place of the LM. The launch was perfect, and the third stage was fired for the first time, to place the CSM into an elliptical orbit. With this success NASA regained much-needed confidence, particularly as the indispensable LM was not only very expensive, but its construction had caused many delays. Consequently when Apollo 5 was launched by a Saturn 1B on 22 January 1968 to test the LM's engines in Earth orbit, the administration was very much on edge. Various problems were discovered as a result of the flight, but none seemed to be insurmountable.

By contrast, Apollo 6 was less successful. The problem lay with Saturn V, which experienced excessive oscillation, causing worrying vibration and the risk of deflecting it from its trajectory. The five second-stage engines also had some defects. However, all the problems had been spotted, and once the causes were known NASA was able to correct them.

Apollo 7 (AS 205) was the first manned Apollo mission. On board the new CSM-101 command module was the old veteran Wally Schirra – who had already flown on both Mercury and Gemini – and two greenhorns, Donn Eisele and Walter Cunningham. As the LM was not needed on this mission, a Saturn 1B rocket was used to put the three men into orbit, on 11 October 1968. The main aim of this mission was to test the reliability of the CSM's SPS (Service Propulsion System) engine by firing it a number of times. The proper functioning of this engine was vital if a crew was to get to the moon and return safely to Earth. Once the Saturn's final stage (the S-IVB) was separated from the CSM, it was to be used for a simulated docking. 14 October saw a notable first: a live televised broadcast. Despite the cramped nature of the command module, the three men found it more comfortable and spacious than Gemini's and they were able to spend 11 days in orbit. The mission was a success, but the astronauts quickly became irritable. They had all caught colds and even though they had a choice of 60 dishes, they found the food unpleasant. What would have seemed a minor inconvenience on Earth became a major burden in their weightless condition. For example, simply blowing one's nose became a major undertaking, as mucus rapidly blocked the nasal passages. The astronauts' irritability had a negative effect on their careers, as none of them was chosen to take part in subsequent missions.

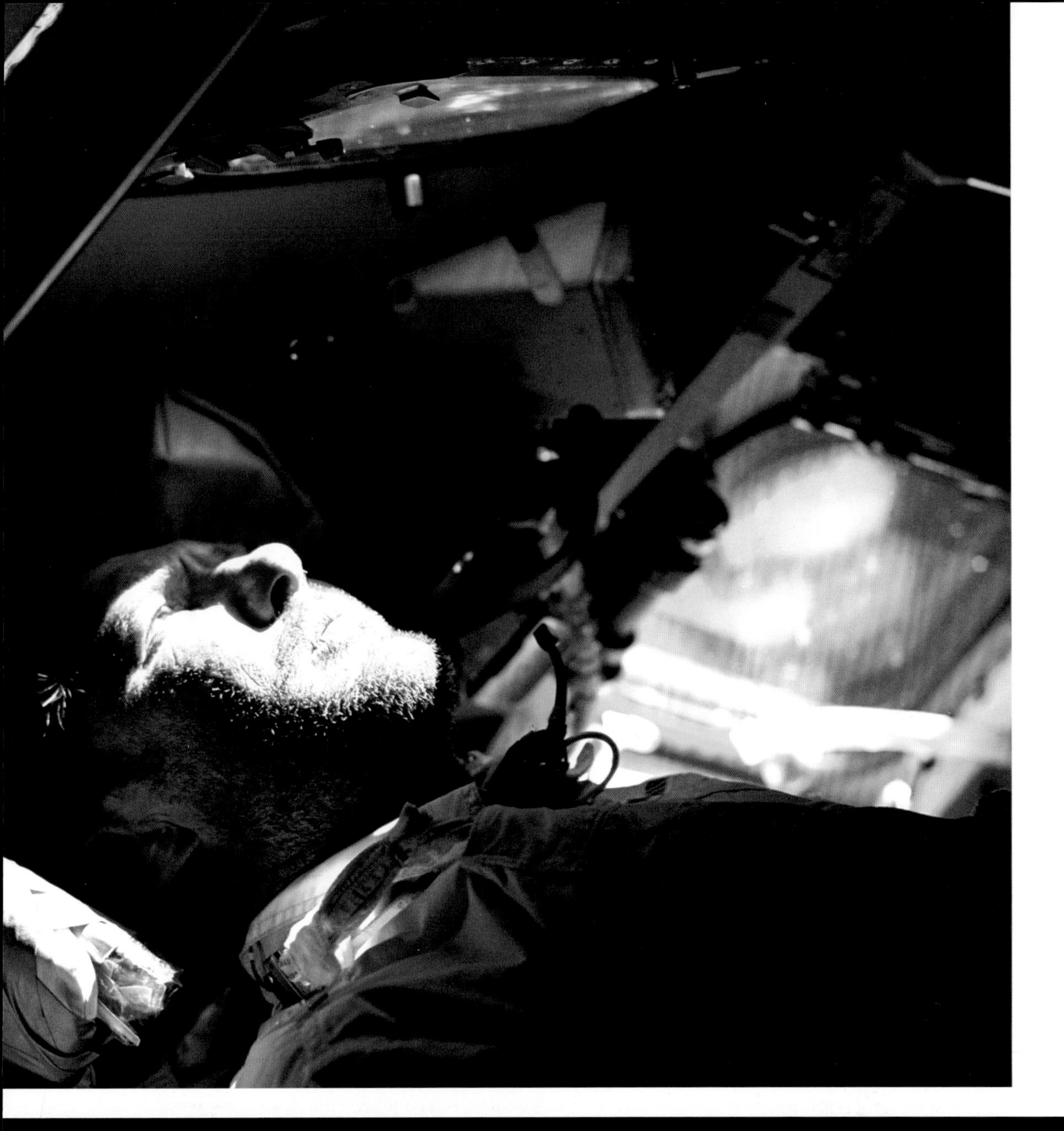

The sun shines on Schirra's face in the command module. Afflicted by a cold, he took Actifed for relief, and for some time was the best ambassador a tablet manufacturer could hope for! The CSM-101 module was from the Block II series that had been thoroughly checked and modified following the Apollo 1 accident.

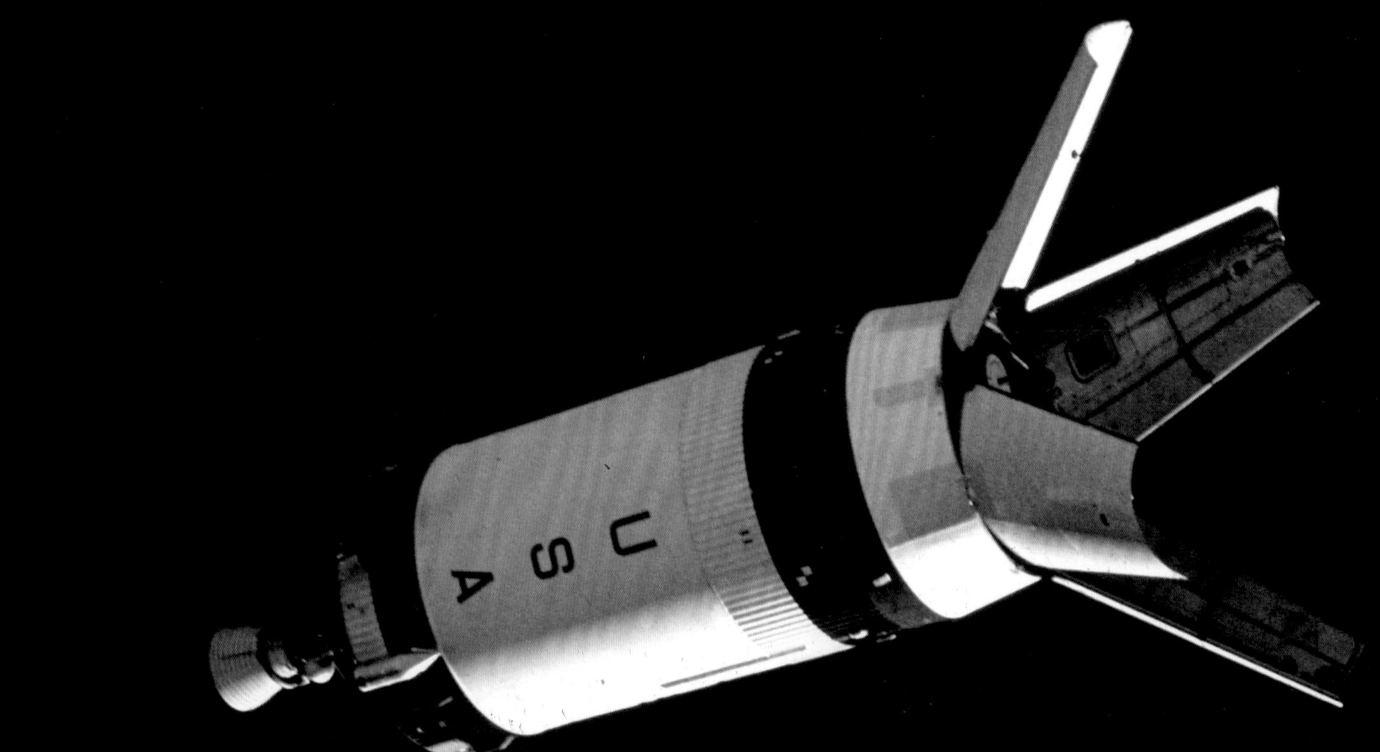

Apollo 7's principal mission was to carry out an orbital rendezvous with the third S-IVB stage of the Saturn rocket, the part that was topped by the CSM and also carried the lunar module. The latter was not used, but it was necessary to be sure that the CSM could detach itself properly from the S-IVB, turn around using its own engines, then dock with the LM. To allow the LM out, the S-IVB's nose opened like the petals of a flower, but during this mission one of the panels did not open fully, though this did not prevent the manoeuvre from being carried out.

Soyuz 3 was equipped with a docking system using a threaded rod that fitted into Soyuz 2. Soyuz 3 was designated 7K-OK BO, and designed solely for docking (it had no airlock at the front). EVAs could be carried out only by opening the main hatch.

A few small leaps for Soyuz

Up until this time little information had been released regarding the early Soyuz missions, but in September 1968 the Soviet Union revealed that an automated craft, Zond-5, had been put into lunar orbit and then returned to Earth. This caused a few concerns amongst the American public, since clearly the Soviets were still on course. The Soyuz 2 and 3 missions caused less of a flutter. Their object was to test the docking of the two craft. For safety reasons, Soyuz 2 was not manned and was launched on 25 October 1968; Soyuz 3, commanded by Georgi Beregovoy, was sent up the following day. The mission was not a success. Placed into Earth orbit, the two craft closed in on each other automatically until less than 700ft away, before Beregovoy took manual control of Soyuz 3 to attempt a closer approach. At less than 150ft the cosmonaut could see the approach lights, of which there were two types to ensure a correctly oriented docking: one set flashing, the other set fixed. Suffering from space sickness, Beregovoy hesitated, preferring to wait until the two craft had moved from shadow into light; but as he waited they drifted apart, and in order to remain in contact with his target he had to repeatedly fire his engines, using much more fuel than had been envisaged. Once the two craft were finally in the light Beregovoy tried in vain to dock, his approach to Soyuz 2 apparently being at the wrong attitude. Eventually he abandoned the attempt, remaining in space for two more days before quitting his 81st orbit to return to Earth.

Soyuz 4 and 5 were put into Earth orbit on 14 and 15 January 1969. They were to dock in space and two of the cosmonauts from Soyuz 5 would go outside and enter Soyuz 4. (At this stage Soyuz was not equipped with a pressurised airlock that would have enabled direct transfer from one craft to another.) Vladimir Shatalov took off alone in Soyuz 4 from Baikonur's Complex 31, followed 24 hours later by Boris Volynov, Alexei Yeliseyev and Yevgeny Khrunov aboard Soyuz 5. After an automatic approach to within 300ft, Shatalov took manual control to achieve a perfect docking, the craft being equipped with a male-female docking system called 'Shtir' on Soyuz 4 and 'Konus' on Soyuz 5. Yeliseyev and Khrunov then donned their Yastreb suits, entered the orbital module, went outside and got into Soyuz 4. Tass declared the success of the operation to be 'the first four-module experimental space station'. Once aboard, the new arrivals removed their suits and symbolically delivered newspapers, telegrams and letters to Shatalov before the craft separated four hours and 35 minutes later. Soyuz 4 returned to Earth with three men aboard, while Soyuz 5 came down carrying only Volynov.

The latter, however, was confronted with a major problem: the service module remained attached to the descent module during

Soyuz 3 took off from the Baikonur launch pad on 26 October 1968. The rocket, taking off only 24 hours after Soyuz 2, is a type-11A511 launcher.

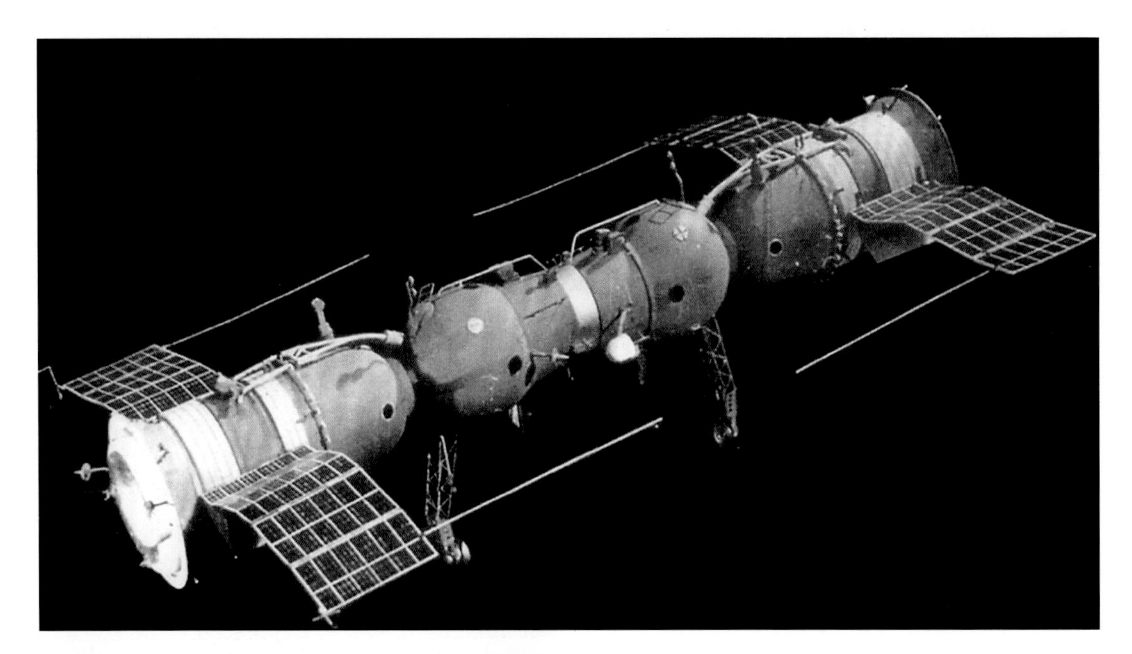

This model shows how Soyuz 4 and 5 were docked. The first craft is the 'male', fitted with a rod that locked into the 'female' docking system. Handrails were fitted so that the cosmonauts could move more easily from one Soyuz to the other.

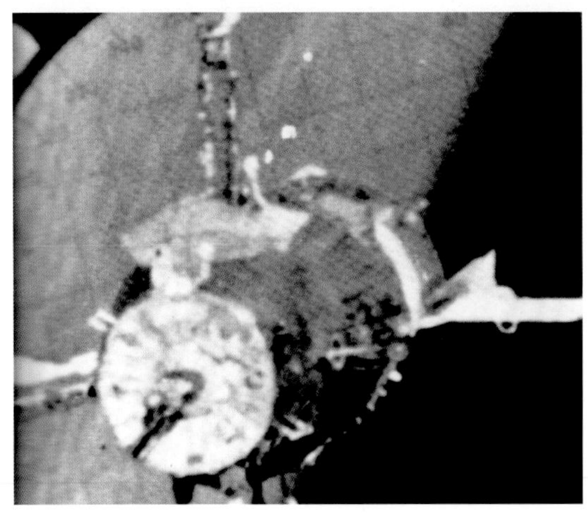

A poor-quality photograph of Soyuz 4 approaching Soyuz 5.

re-entry, after firing of the retro-rockets. On Soyuz this could have very serious consequences, as, with the two craft joined, the descent module was positioned with its nose pointing forwards, and this lacked a proper heat shield. (The normal procedure was for the module to turn around, leaving the rear section, clear of the service module, to serve as a heat shield.) Very soon smoke from the atmospheric friction began to seep in through the access hatch and Volynov began to find himself in a somewhat uncomfortable situation, with the centrifugal force having a tendency to push him out of his harness instead of holding him down. Then a miracle occurred: the heat burned through the straps holding the two craft together, the service module violently tore itself away and the descent module took up its correct attitude. However, Volynov was not yet out of the woods, as the main parachute had difficulty in opening fully and the thrusters intended to reduce the speed of descent immediately before touchdown did not work. *Soyuz 5* therefore hit the ground more heavily than it ought to have done and Volynov ended up with broken teeth. But he was alive...

Celebrated as heroes, the four Soviet cosmonauts could not quite hide their disappointment. Even though, in terms of quantity, the Soviet Union had sent a lot more Soyuz craft into space than the total of Apollos sent up by the Americans, the difference was that an American crew had flown right around the moon in December 1968, and NASA had thereby established a significant lead.

Beregovoy is in the centre, in uniform. He was unable to dock properly with Soyuz 2. At this time the manoeuvre was carried out manually, but the way the craft was positioned prevented the sensors from working correctly. The excessive amount of fuel that was used up forced Beregovoy to abandon the mission. This former Stormovik pilot and Hero of the Soviet Union was getting on a bit (he was 47) at the time of the flight, but someone with experience was required. While he made mistakes during the flight, it should be pointed out that he was on his own and Soyuz was a much less manoeuvrable craft than, for example, Apollo.

Yeliseyev (left) and Khrunov (right) left Soyuz more or less simultaneously to transfer to Soyuz 4, where Shatalov was awaiting them. No problems were encountered during the operation. Shatalov did not need to put on a spacesuit, as he remained within the pressurised descent module.

This picture of the Earth rising over the moon has become famous. Nobody had ever seen the Earth like this. The photograph was taken on Apollo 8's fourth orbit of the moon. Public opinion was so aroused that *Time* magazine named the three astronauts 'Men of the Year'. Borman felt inspired to quote a few lines from Genesis, which caused a bit of a stir when a noted American atheist brought a suit against NASA for promoting Catholicism in a programme paid for with public money!

The moon within reach...

The Apollo 8 mission marked a decisive point in the race to the moon: for the first time, a crew (consisting of Borman, Lovell and Anders) left Earth orbit and flew to the moon. It was also the first manned launch by a Saturn rocket, and the mission's complete success gave NASA renewed confidence.

However, not everything was rosy for the administration. The timetable set down in 1967 had envisaged several flights with the LM, or Lunar Module, but the Grumman Company who had been given the job of designing it had fallen behind schedule and did not hand it over until June 1968. No fewer than 101 faults were detected on this 'pre-LM', giving rise to a notable dispute between NASA's managers and the Grumman engineers, who were obliged to put right all the faults so as to make it operational. Under these circumstances, it

The firing of a Saturn V rocket was very impressive, its thrust being in the region of 3,350 tons! The F-1 engines were supplied from a 175,000-gallon kerosene tank topped by a 250,000-gallon liquid oxygen tank. The whole of this mixture was used up in about two and a half minutes, the time it took to reach an altitude of about 40 miles.

Jim Lovell, Bill Anders and Frank Borman encountered no problems.

Lit up like a Christmas tree, the Saturn V rocket quietly awaits the countdown to end before bursting into life. The Apollo 8 mission was of fundamental importance not only to NASA, but also to most of mankind. For the first time ever, men were flying to the moon, circling it and then returning to Earth. The fact that NASA prepared the mission in under four months was all the more remarkable for the fact that it was only the second manned mission of the programme.

was clear that the Apollo programme would suffer delays.

It was at this juncture that George Low, the head of Apollo, took a decision: in order to test the concept of the CSM in lunar orbit, why not send a module directly to the moon? This would not only be a step onward from Apollo 7, but would also match the Soviets, whom the CIA suspected of preparing a lunar-orbit mission – two cosmonauts, Leonov and Makarov, had been specially trained to embark on a Soyuz craft. This choice worried James Webb, NASA's administrator, as without the LM the crew had no 'life-raft' (in the event of an engine failure on the CSM, the LM's engines could have been used instead). Over the next few months the programme was revised, temporarily dropping missions D and E – intended to put CSM and LM modules into a low orbit followed by an elliptical orbit – and going straight for a lunar-orbit mission.

The crew of Apollo 9 were to test docking and undocking manoeuvres with the lunar module while remaining in low Earth orbit. The mission was a complete success and lifted spirits after the delays with the LM. The commander was Jim McDivitt (right), a former Edwards test pilot, who had already commanded the Gemini 4 mission. After Apollo 9 he would play an important role in the continuation of the lunar programme when he became director of lunar landing operations, then directed the Apollo craft programme as far as Apollo 16. David R. Scott (centre) was the pilot of the CMP module. He had also flown before, having his space baptism on Gemini 8. Rusty Schweickart (left) piloted the lunar module. He suffered from space sickness and would not fly again. Here, on 3 March 1969, the three men are about to set out on their ten-day adventure.

Men around the moon

Announced in December 1967, the Apollo 8 mission started a year later to the day. In the meantime, Lovell had replaced Collins, who had to undergo surgery. The launch passed off without incident and, once they were in orbit, the crew checked all the systems before entering TLI (Trans-Lunar Injection) mode, which would propel Apollo 8 towards the moon. The S-IVB stage of the Saturn rocket was fired for five minutes and 17 seconds before being jettisoned at an altitude of 215 miles. The craft's speed now reached more than 6.8 miles per second and the command module was turned so that a few photographs of the Earth could be taken – the first pictures of the entire planet ever taken by a human being. As they sped away, Borman, Lovell and Anders passed through the Van Allen Belt – a radiation belt discovered by James Alfred Van Allen at the

This interesting photograph shows how the LM appeared once the CSM was separated from the Saturn S-IVB stage at an altitude of around 120 miles. The lunar module is still nestling in the upper part of the S-IVB and the CSM will shortly close in to dock. All this took place within three hours from lift-off. When the two craft were properly linked up, the LM and CSM were gently ejected from the S-IVB stage, which then fired its engines and moved away.

end of the 1950s – whose effects on mankind were little known. Now, more than 15,000 miles from Earth, the crew of Apollo 8 proved that the radiation was harmless.

After this, the spacecraft went into PTC (Passive Thermal Control) mode, also known as 'barbecue' mode: in space, anything in full sunlight could reach a temperature of 200°C, while objects in the shade could drop to −100°C. To spread these temperatures evenly, the spacecraft rotated once every hour. A few more engine burns allowed correction of the trajectory and the crew settled down to periods of sleep and wakefulness. When Borman woke up, he felt sick and even vomited, although the medical team on the ground did not become unduly worried, as the problem got no worse. Thirty-one hours into the flight a television link was set up, in which Lovell sent birthday greetings to his mother. Condensation on the window prevented the crew from seeing the moon, but a second broadcast gave audiences their first TV views of the Earth from space. The moon's gravitational pull began to be felt 55 hours after launch, necessitating a slowing of the CSM from 4,000ft/sec to 2ft/sec.

Less than 17 days later, Apollo 8 got ready to go into lunar orbit. The craft would then pass behind the dark side of the moon, cutting all communication with the Earth. Entry into orbit was a delicate manoeuvre, as the RCS's engine had to fire for exactly the right length of time. Too short a burn would have sent Apollo 8 into an elliptical orbit with the risk of being hurled out into space. A burn of just a second too long might have set the craft on a collision course with the moon... Emotions were thus running high, aboard Apollo 8 as well as back on Earth.

When the module reappeared from behind the moon it was in the right position and dead on time. Lovell was able to describe the lunar surface in greater detail than had ever before been possible: 'The moon is essentially grey...like plaster of Paris or a sort of greyish sand...' The camera took pictures of the Sea of Tranquillity, where it was planned to land Apollo 11. Borman then read a passage from Genesis, which was transmitted to his church. Later, during the fourth orbit of the moon, the crew took one of the finest pictures ever taken in space: Earth rising over the moon.

After a rest and another TV transmission, it was time for the crew to head back to Earth. The TEI (Trans-Earth Injection) required another engine burn, which went perfectly and allowed the module to get into the correct trajectory for its return to Earth. 'Please be informed there is a Santa Claus,' claimed Lovell. 'That's affirmative,' replied Capcom back on Earth, 'you're the best ones to know.' It was 25 December 1968 and Apollo 8 had fulfilled its mission, having collected all the information needed for a flight to and a landing on the moon. The return flight was a model of precision and the crew completed their journey with a splashdown in the middle of the Pacific. At the conclusion of this historic flight, President Johnson declared: 'You have taken all of us, all over the world, into a new era.'

Several sorties into space were undertaken, which provided an opportunity to test the new spacesuit, with its independent Personal Life Support System (PLSS), for the first time. Scott is here emerging from Apollo 9 to take a photograph of Schweickart. It was on this mission that NASA once again authorised the use of nicknames for the craft. The LM was called 'Spider' and the command module 'Gumdrop' (because the plastic covering over the craft before its launch looked like chewing-gum wrapper foil).

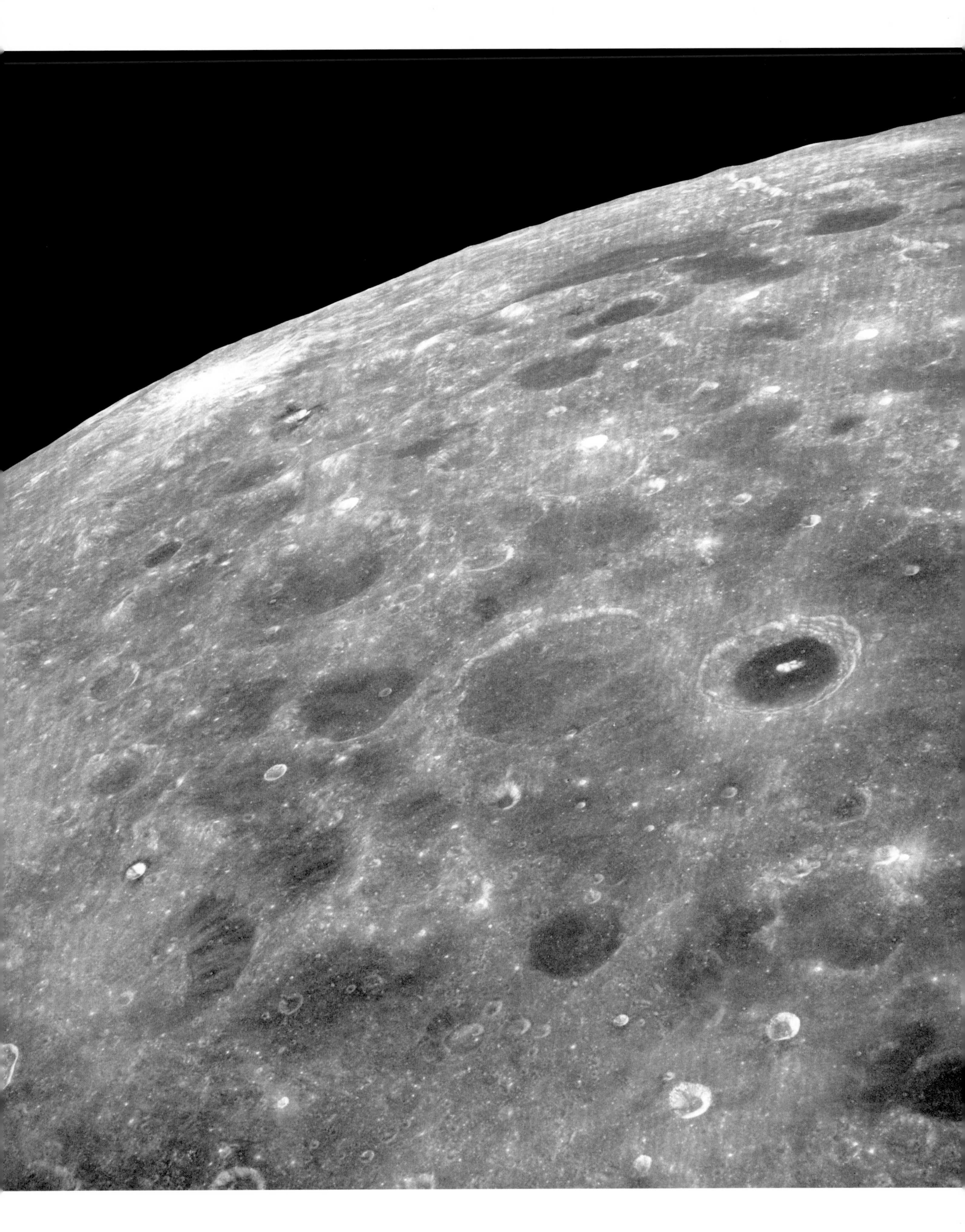

Wearing his PLSS on his back, Schweickart carried out an EVA of just over an hour. The equipment weighed 57lb, to which had to be added the 50lb of the spacesuit. Obviously, the lack of gravity considerably reduced this.

Opposite: 'Spider' moves away from the CSM with McDivitt and Schweickart on board. The latter was continually sick, but was able to carry on until the end of the mission. The LM moved quite some distance away from 'Gumdrop'. This exercise was very risky, because if the engine failed to work McDivitt and Schweickart had no means of getting back.

At last, the lunar module enters the scene

NASA had now achieved its first objective: to send a team around the moon and bring it back safely. Another reason the administration could breathe a little easier was that it seemed fairly clear that the Soviet Union was not going to be able to send a man to the moon for some years hence: the big N1 rockets, the Soviet counterpart of the Saturn, were all deficient in some way. It only remained to put a man down on the moon's surface, which was exactly what was required of the Lunar Module, or LM, a complex craft with the appearance of a delicate insect. But before attempting this tricky mission, NASA would have to be sure that the two-man Command and Service Module (CSM) and Lunar Module would work. This trial was carried out on the Apollo 9 mission. James McDivitt and David Scott – two veterans of Gemini – along with Russell Schweickart lifted off from the Kennedy Space Center's launch pad 39A on 3 March 1969 on a Saturn V rocket. Its task was to prove the CSM and LM in Earth orbit. The CSM separated from the S-IVB stage less than three hours after insertion into orbit, then turned around ready to dock with the LM, now nicknamed 'Spider'. The LM was located on top of the S-IVB in a compartment that opened out like the petals of a flower.

The crew practised for ten days. Schweickart and Scott went out into space wearing the new Apollo suits (the first truly self-contained suits, with no umbilical cord supplying air and cooling water). Eventually, McDivitt and Schweickart entered the LM, undocked from the CSM (with Scott on board), and moved until they were about 125 miles away, before returning to re-dock. The crew re-formed on board the CSM and returned to Earth on 13 March. Once again, the mission was a complete success. A little more than two months later, Apollo 10 would be a full-scale rehearsal for Apollo 11, omitting only the lunar landing. Put into a 'parking' orbit above the Earth, the S-IVB stage, topped by the LM and CSM, was fired to take Thomas Stafford, John W. Young and Eugene Cernan to the moon. The CSM detached itself from the S-IVB, carried out its turning manoeuvre and docked with the LM before jettisoning the S-IVB. With the latter no longer required, it was sent off into a solar orbit and the combined CSM-LM, weighing 42 tons, continued its journey. The mission then proceeded in the same way as Apollo 8. Once into lunar orbit, Cernan entered the LM (this time nicknamed 'Snoopy') to check the state of the vehicle. On 22 May, the LM separated from the CSM ('Charlie Brown') with Stafford and Cernan aboard and descended to an altitude of about ten miles above the intended landing site of Apollo 11. The descent engine worked properly, and the LM then separated from its descent stage and came back up again using its ascent engine. The re-docking with the CSM that had remained in orbit went perfectly and the two men rejoined the CSM. Before setting off for Earth, the LM was jettisoned into a solar orbit and remains to this day the only intact LM, the others having either crashed on the moon or burnt up on re-entry into the Earth's atmosphere. On 26 May, Apollo 10 came down in the Pacific, just three and a half miles from its intended spot.

'Gumdrop' poses for its portrait, taken from the lunar module. Clearly visible are the communications antennae (still pointing towards the Earth). The large rectangular porthole visible at lower right was used for observation. Two other recessed portholes were used during rendezvous manoeuvres.

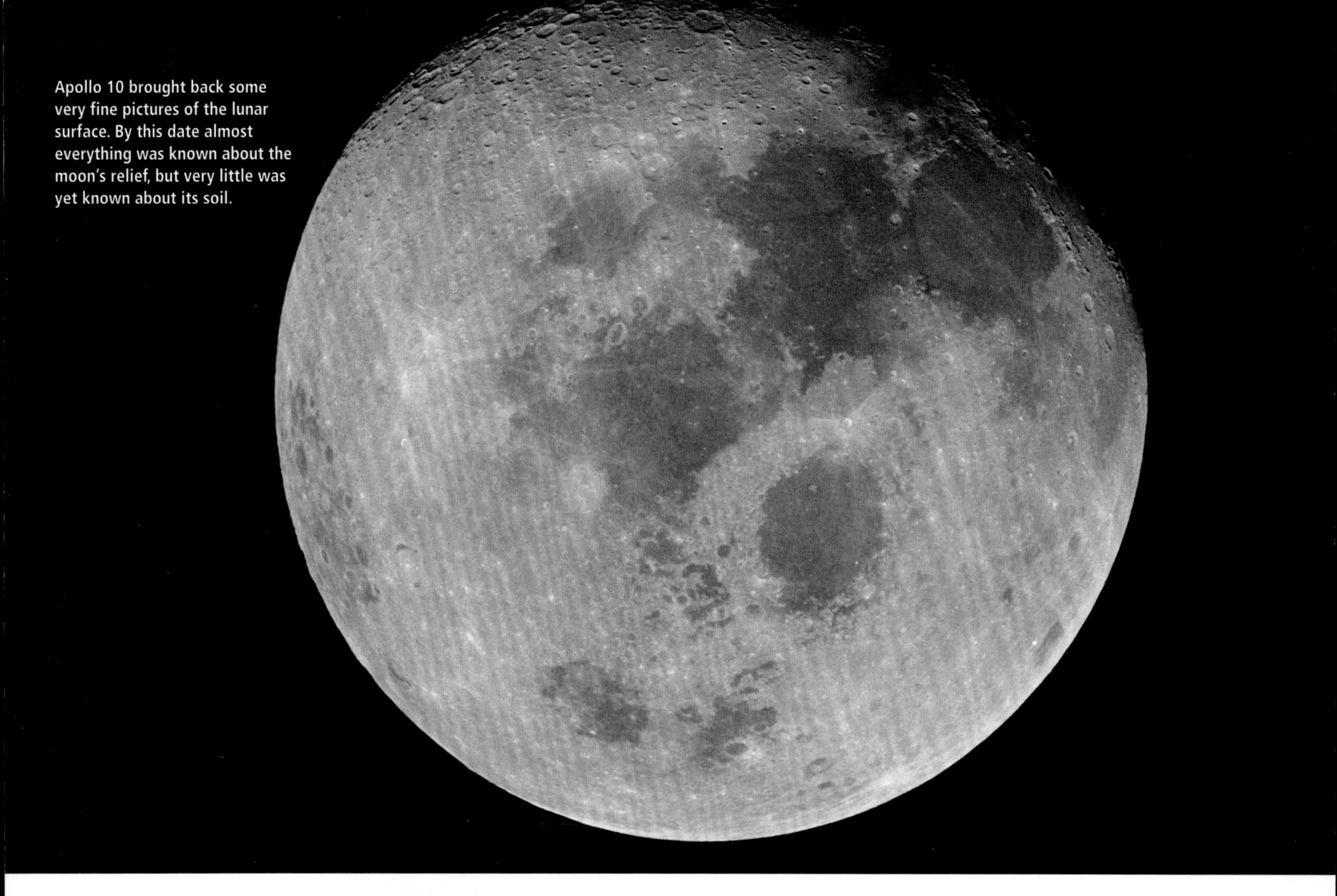

Apollo 10 brought back some very fine pictures of the lunar surface. By this date almost everything was known about the moon's relief, but very little was yet known about its soil.

The veterans: Cernan, Stafford and Young. All three had previously been into space on Gemini. On the way to the bus taking them to the launch pad, the mission's mascot, Snoopy, gets a friendly pat. 'Snoopy' was also the name given to the lunar module, with the command module being known as 'Charlie Brown'.

Donald Kent 'Deke' Slayton, confined to the ground since the Mercury missions, effectively directed the crews' flights between 1963 and 1973 before becoming one of the Apollo-Soyuz astronauts. He is seen here with Charles Moss Duke, who was to become the most famous voice of Capcom during Apollo 11. He would get his turn to go to the moon on Apollo 16. The tension can be seen in the two men's faces.

The lunar module slowly moves away from the CSM in orbit above the moon, with Cernan and Stafford on board. The 'Charlie Brown' module was not equipped to land on the moon, and was limited to flying over the future Apollo 11 landing zone, the Sea of Tranquillity.

Walking on the moon!

Apollo 11 was the most spectacular mission and the one that remains in the public's collective memory as the high point in the conquest of space. With this mission, NASA fulfilled President Kennedy's promise of 1961 with just months to spare. Kennedy's desire to send a man to the moon was never questioned by either of those who succeeded him in office (Lyndon Johnson and Richard Nixon), and the Apollo 11 mission became the realisation of a national dream; but as soon as the dream was achieved the federal government made considerable cuts in NASA's budget. On 16 July 1969, however, all eyes were turned towards the huge Saturn rocket sitting on its Florida launch pad. A few days earlier, on 3 July, the Soviets had attempted a final launch of an

On 16 July 1969, at 09:32 local time, the Kennedy Space Center's launch pad 39 once again experienced the deafening roar of a Saturn V. The rocket seen here would help the Americans to achieve President Kennedy's objective of sending men to walk on the moon.

In the firing room, von Braun is delighted, while outside Lyndon Johnson manages a slightly frozen half-smile, his eyes fixed on the rocket lifting off.

unmanned Soyuz 7K-L1A-type craft using an N1 rocket. A quarter of a second after ignition, one of the engines suffered a failure and the rocket exploded, destroying a large part of the launch pad. For the Soviets, it was effectively the end of the race to the moon...

The Apollo 11 crew consisted of three veterans of space flight, all of them having taken part in the Gemini programme: Neil Armstrong was the commander, Michael Collins the pilot of the command module ('Columbia'), and Edwin 'Buzz' Aldrin the pilot of the lunar module ('Eagle'). At 09:32 local time, the five F-1 engines roared into life and the rocket began to rise slowly skywards. Regular as clockwork, it propelled its crew into orbit, circling the Earth

Neil Armstrong (CDR), Michael Collins (CMP) and Buzz Aldrin (LMP) are without question the best-known astronauts in the world. Yet Collins would never set foot on the moon. In 1970, he left NASA and a year later became the first director of the NASM (National Air and Space Museum) at the Smithsonian in Washington DC, not far from the Capitol.

On 19 July 1969 the CSM and the LM were attached to one another in a configuration known as a 'trans-lunar orbit'. The two vessels flew again over the intended landing site. In this picture Armstrong has captured Aldrin getting ready in the lunar module, 'Eagle'. The LM separated easily from the CSM, thanks to a spring system that allowed the LM to achieve a speed of a metre a second. The LM then stopped at around 50ft from the CSM so that the command module's pilot could visually check that the LM's landing legs had deployed correctly. An hour and a half after the separation of the two vessels, the LM started its 50,000ft descent to the moon's surface. This consisted of three phases: braking, approach and final touchdown.

one and a half times. The first part of the mission was identical to that of Apollo 10. On 19 July they went into lunar orbit. Having inspected 'Eagle', Armstrong and Aldrin took over the controls and separated from the CSM, where Collins remained. The LM stayed for a few minutes above the lunar surface while Collins inspected it from every angle. When nothing out of the ordinary was detected, the LM started its descent engine. Very soon, Armstrong and Aldrin realised that they were going to touch down 'too long', their rather too rapid descent likely to set them down some miles from the planned location. Suddenly an on-board alarm rang out and a red light began to flash on the D-SKY (Display and Keyboard, the LM's computer control system). The light was a 1201 alert, indicating a major problem. In fact, the central computer could not handle all the information it was receiving, because the radar guidance between the LM and the CSM had not been disconnected and was overloading the LGC (the LM's central computer) with useless data. On the ground,

Armstrong photographed Aldrin coming down out of the LM. The manoeuvre was not that easy and he had to half-kneel, restricted in his spacesuit, slide onto a platform and grab the ladder before he could finally get down. The module remains to this day a real technical achievement. At 14.5 tons it was quite a lightweight. Everything had been done to save a few pounds. Thus while the gold-leaf coating protecting the descent stage (the lower part of the LM) had certainly been expensive, it allowed a saving of 110lb over the Mylar and aluminium thermal protection that was also considered.

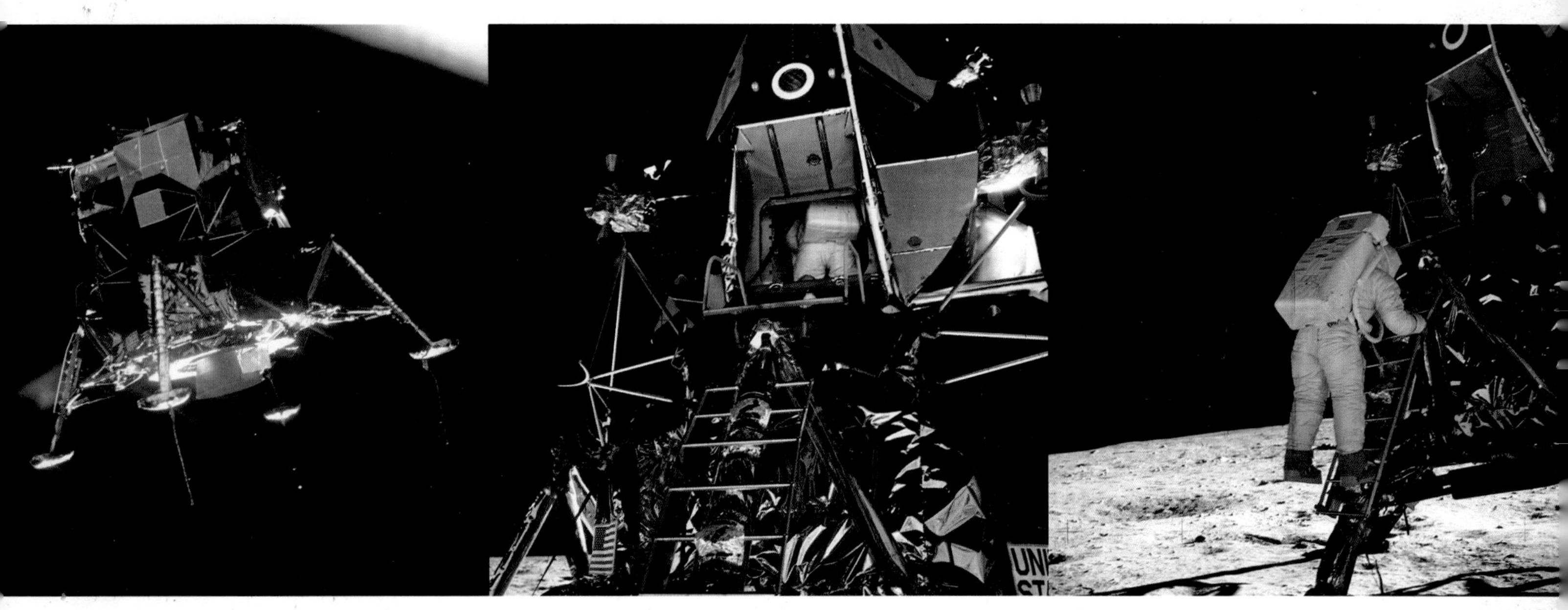

The first human footprint on the moon, photographed by Armstrong. The soil turned out to be dustier than had been expected, 'like talc' according to Armstrong. This grey dust got everywhere, and on the ensuing missions – where a considerable amount of material was being picked up from the lunar surface – you could see the astronauts' suits were completely covered in a film of it. One of the first things the two men did was to erect an American flag, which had to be supported, as there was no wind for it to flap in!

thoughts turned to aborting the mission, as fuel was being consumed too fast and it was vital to maintain a reserve of 20 seconds in case the mission had to be abandoned. At Houston mission control, concern was at its height. After a quick consultation, a 25-year-old engineer, Steve Bales, took the decision to continue the descent.

Armstrong closed down the automatic control and took over manually, at the same time glancing outside to see that their landing site was a crater with a much rockier-looking surface than had been expected! At 16:17 (Kennedy Space Center time), 'Eagle' touched down on the Sea of Tranquillity. There were only 30 seconds of fuel left for the DPS (Descent Propulsion System)... It was Aldrin who spoke the first words: 'Contact light,' meaning that a light indicating touchdown had illuminated on the control panel. Armstrong then came out with his famous: 'Houston, Tranquillity base here. The Eagle has landed.' Charlie Duke, director of Capcom in Houston, let out a sigh of relief: 'We're breathing again. You got a bunch of guys about to turn blue.'

Another six and a half hours passed before Armstrong put on his spacesuit, descended 'Eagle's' ladder and put his foot down on the lunar surface. He then spoke the words that went around the world, broadcast live on radio and television, watched by between 600 million and a billion people: 'That's one small step for [a] man, but one giant leap for mankind.' It was 10:56. After taking a few pictures, he was quickly joined by Aldrin. They set up an American flag and President Nixon was able to speak directly to the two men. They then set to work using the 'Early Apollo Scientific Experiment Package (EASEP), a laboratory that was designed to carry out a variety of experiments (a laser reflector with a beam pointing from Earth allowed them to

measure as accurately as possible the distance from Earth to the moon). A movie camera was set up and sent back black-and-white pictures of the two astronauts moving around with more ease than they had experienced under simulation. By contrast, getting into and out of the LM was tricky because of the size of the PLSS (Personal Life Support System), the backpack containing the spacesuit's oxygen and cooling and communications systems. They gathered some 50lb of moon rocks then, after two and a half hours of activity, they got back on board the LM. They got rid of their PLSS, over-boots and a Hasselblad camera. At 1:54, they fired the return module's engine and rejoined the CSM in orbit. The LM was jettisoned and the crew began the return to Earth. They got back on 24 July, but, because of the risk of having brought back bacteria from the moon's surface, they were kept in quarantine. President Kennedy's aim had been achieved, with just five months remaining to the end of the decade…

The famous picture of Aldrin on the moon. The A7L-type suit worn on the lunar missions was fairly complex, but it gave complete satisfaction. Not only did it provide the oxygen needed for life support, but it also gave protection against meteorite dust, cold and heat. The gold-tinted visor protected the eyes from ultraviolet light and the sun's rays. The helmet was covered in a kind of hood that gave heat protection around the neck. Being exposed to the sun in the lunar vacuum was extremely dangerous, as there was no atmosphere to moderate its heat.

On early Apollo missions the crews were quarantined afterwards, just in case… Eventually, however, this precaution was dropped. Here Richard Nixon visits the three heroes, who are feted right across the country.

Three of the seven cosmonauts sent almost simultaneously into space in October 1969: Shonin, Filipchenko and Shatalov. After giving up on the moon, the Soviets sought new ways of getting their space programme back on track. They therefore tried sending up three Soyuz craft. Two of them were to rendezvous, with their crews doing EVAs, while the third was to remain close by and film the operation. The launches and meeting up of the craft were trouble free, but once again the docking system failed.

Nikolayev and Sevastianov, the two heroes of the Soyuz 9 mission, were in some respects pathfinders for the future Salyut missions. The amount of work they were asked to do was enormous and both returned to Earth in a state of complete exhaustion.

The Russians quit the race...

Now left far behind in the race to the moon, the Soviets concentrated their efforts on rendezvous in space, which were initially unsuccessful. In October 1969 they launched no fewer than three Soyuz (6, 7, and 8), putting a total of seven cosmonauts into space simultaneously. Three Soyuz 11A511 derived from the R7 were launched from Baikonur on 11, 12 and 13 of October. The intention was for Soyuz 7 and Soyuz 8 to dock, with the manoeuvre being filmed from Soyuz 6. Two cosmonauts from the first craft would then transfer to Soyuz 8. Various other experiments were to be carried out, notably a welding test under weightless conditions using the Vulkan system. However, for some reason not clearly understood the docking manoeuvre failed, effectively putting an end to the Soviets' lunar pretensions. Nonetheless, the three craft remained in orbit for five days before returning to Earth without any problems.

Very soon, the Soviets turned their attention towards orbiting space stations, or, initially at any rate, carried out long-duration flights, partly to wrest the record from the Americans and also to test the reaction of the human body to periods of weightlessness of between 17 and 20 days.

It was with this aim in mind that Soyuz 9 was sent up on 1 June 1970. The launch was to have taken place on 22 April 1970, the 100th anniversary of Stalin's birth, but the scientists did not have the programme ready in time and the launch was postponed several times. Soyuz had not originally been designed for flights as long as 20 days, so the craft was at the limit of its capabilities. Andrian Nikolayev and Vitali Sevastianov lifted off without any problems. Nikolayev was undoubtedly one of the best of the Soviet cosmonauts. Coming from the first group of 20 candidates selected in 1960, he had already flown on Vostok 3 in August 1962. Of a naturally calm disposition, he had married Valentina Tereshkova in November 1963. (In fact the marriage had been arranged by the Supreme Soviet, which was more than happy to use the two space heroes for propaganda purposes. The couple divorced shortly afterwards.)

Once in terrestrial orbit, Soyuz 9 started to develop problems. On the fourth day, one of the solar panels began to show a severely reduced power output, obliging the crew to rotate the craft so that the offending panel would be regularly exposed to the sun.

As the days went by, the condition of the cosmonauts seemed to deteriorate. They made a few inconsequential mistakes, but their concentration level began to drop. At the same time, carbon dioxide levels started to grow to a potentially dangerous level. By the 17th day the Soviets had beaten the space endurance record, and on 19 June Soyuz 9 returned safely to Earth. However, its crew were in such a poor state that they could not walk and had to be carried during their return to Moscow. Everyone had thought that remaining in a state of zero gravity would have no effect on the metabolism of the human body. Mishin, Korolev's successor at the head of OKB-1, even thought that a man could stay in space for a month or two without appreciable ill-effects. It therefore required all the efforts of the leader of the cosmonauts' group, Nikolai Kamanin, to have Nikolayev and Sevastianov put under immediate medical surveillance, especially as the party bosses wanted them to be available for a plethora of celebrations. Even ten days after their flight, the men could concentrate for only three or four hours a day, such was their state of total exhaustion.

Pete Conrad, Richard Gordon and Alan Bean made up the Apollo 12 crew. Conrad was one of the most experienced of the American astronauts. An outstanding military pilot, he had been selected for the Mercury programme and was later allocated to the Gemini missions. He took part in two of them (Gemini 5 and Gemini 11) before commanding Apollo 12. Gordon had already been into space with Conrad on Gemini 11. The two men got on well and knew each other before joining NASA. The last of the trio, Alan Bean, had also worked alongside Pete Conrad, as the latter had been his instructor when they were Navy test pilots. This was indeed a closely-knit team.

Achievement becomes routine

The Americans meet up with Surveyor 3

Without the general public being aware of it, Apollo 11 effectively marked the end of the Apollo programme, although the ensuing missions were certainly worthy of interest. The race to the moon, though, was well and truly over, with no further competition in sight. The other missions were thought to be rather less exciting and NASA's budget was cut substantially. The 'Great Society' programme launched by President Johnson – who nevertheless broadly supported Apollo – and the Vietnam War, continued by his successor Richard Nixon, were to significantly reduce NASA's budget from 1966 onwards. Thomas O. Paine, who succeeded James Webb as NASA's administrator, put forward the idea of a space aeroplane that could be re-used, thus cutting costs and increasing productivity, an idea likely to appeal to the White House, although it was a long way from realisation.

While awaiting the Shuttle, the Apollo programme continued... Apollo 12's mission began on 14 November 1969 under poor weather conditions with Pete Conrad, Richard Gordon and Alan Bean on board the command module, 'Yankee Clipper'. The LM, 'Intrepid', was to make a moon landing near the Sea of Storms, where a number of probes such as Luna 5, Surveyor 3 and Ranger 7 had come down. Except for an incident on lift-off when a lightning bolt struck the command module and cut the electric power supply for a few seconds, the Apollo 12 mission was a model of precision. The LM needed almost no manual course correction and made a soft landing at exactly the right spot, less than 700ft from Surveyor 3. Conrad was of a smaller stature than Armstrong and had to jump onto the moon's surface from the final rung of the ladder, shouting: 'Man! That may have been a small step for Neil, but that's a long one for me!' He was joined half an hour later by Bean, and the two men started their experiments. Apart from the now standard Stars and Stripes flag, the astronauts set up a new atomic-powered, mini-laboratory. After going back on board the LM, they carried out a second EVA, removing some parts from Surveyor 3 to take back to Earth, and then collecting samples from the moon's surface. In all, the two EVAs lasted nearly four hours each out of a total

Apollo 12 was one of those near perfect missions...except for the lift-off. The weather had been bad, with squally rain as can be seen in this photograph of the lift-off showing raindrops splashed on the lens. Then just after take-off the rocket was struck by lightning. Strangely, it ran down the rocket's body and then along the road taken by Saturn V as far as the tower! All the on-board electricity was briefly cut, and the alarms signalled numerous system failures. In the firing room, everyone was even more concerned by the fact that no information from the rocket was reaching them. Gerry Griffith, the flight director, considered aborting the mission, but the EECOM (Environmental, Electrical and Instrumentation Manager) gave the order to switch to 'SCE aux'. No-one in the firing room or on Apollo had ever heard of this button! Suddenly, Bean remembered that he had used it during a simulation a year before. When they went over to the auxiliary supply all the systems were reset and the mission was able to continue.

The lunar module gets ready to carry out its descent manoeuvre with Conrad and Bean. Apollo 12 allowed NASA to make much more precise lunar touchdowns. Bean was the first to spot the Surveyor 3 probe launched in April 1967, which had taken pictures of potential landing sites for the Apollo LMs.

Conrad and Bean carried out many scientific experiments, collected samples, set up the ALSEP and recovered Surveyor 3's camera so that it could be taken back to Earth. They spent a total of seven hours and 45 minutes on their walkabouts.

time on the moon of 31 hours and 31 minutes. They had taken a colour movie camera, but Bean mishandled it and it failed to work properly. The two men rejoined the CSM, jettisoning the no-longer-needed LM, which crashed onto the moon's surface on 20 November, registering on the seismographic recorders. They and Gordon remained a further day in orbit to complete the photographic record. On the return flight, some time was devoted to interviews and by 24 November the crew was safely back.

A tragedy avoided

History shows that a defeat can be transformed into a victory by heroism. This is exactly what happened with the Apollo 13 mission. Apollo 12 had been a perfect mission, perhaps too much so, with the result that everyone thought that the ensuing one would be likewise. The tragedy of Apollo 1 had been forgotten, the more so as it had taken place on the ground and NASA's engineers had dealt with its causes.

The Apollo 13 mission commander, James Lovell, was an experienced astronaut – he had flown on Gemini 7, Gemini 8 and Apollo 8 – but this time, the number 13 would bring him bad luck... Accompanied by John Swigert – replacing Ken Mattingly, who had been chosen as commander of the module but had had to withdraw at the last minute when he contracted rubella – and Fred Haise, he boarded the command module, 'Odyssey', on 11 April 1970. It was a more or less routine launch, with the by-now familiar stages: the Saturn V rocket lifted off from launch pad 39 at 14:13, the S-IVB final stage went into stationary orbit a few minutes later. and less than an hour after lift-off the J-2 engine was fired, sending the modules into a trans-lunar trajectory. 'Odyssey' uncoupled, manoeuvred into position and docked with 'Aquarius', the lunar module. The S-IVB stage fired its engines again to take itself away, crashing into the moon three days later at around 850ft/sec. The violence of the impact was felt by the seismograph left by the Apollo 12 mission.

James A. Lovell (CDR), John L. Swigert (CMP) and Fred Haise (LMP) made up the crew of Apollo 13. Lovell was the most experienced, having three flights on his CV, including one on Apollo 8. Swigert saw his career cut short by his role in the Apollo 15 'stamp scandal' (in which he was not directly implicated). Haise had been the backup for a number of Apollo crews. He was to have been the mission commander on Apollo 19 until this was ultimately abandoned. He later worked on the Space Shuttle project and flew some approach and landing test flights on *Enterprise*.

Up to this point, everything had gone as normal and the crew put 'Odyssey-Aquarius' into exactly the right trajectory to orbit the moon at an altitude of 70 miles. On 13 April 1970, at 13:13 – this could not be made up – and 200,000 miles from the Earth, Apollo 13 was asked to stir the oxygen in the tanks, which had a tendency to freeze on the walls, impeding the sensors and giving false pressure readings. In short, it was a routine operation, a housekeeping task... At that moment, a muffled explosion was heard. For a second, Lovell and Swigert thought it was just another of Haise's innumerable pranks – he occasionally opened a relief valve to scare his colleagues. At the time, Swigert was in the command module, Haise in the tunnel leading to the lunar module and Lovell halfway between the two. When Lovell saw Swigert's head, he knew something was wrong. An alarm went off, alerting the crew that the on-board electric power was about to give up the ghost because of the failure of the fuel cells, which needed oxygen to work. The situation was critical, as the capsule could rely on its own batteries for only ten hours and they needed at least 24 hours to get back to Earth; furthermore, they needed to conserve battery power for the re-entry into the atmosphere after separating from the service module.

'Hey, Houston, we've had a problem!' said Swigert – not Lovell as is often thought. Houston's Mission Control had also noticed

The lift-off was perfect, but the middle engine in the second stage cut out earlier than it should have done, which was fortunately compensated for by the extra power of the other engines. Apollo 13's mission was to explore the Fra Mauro Mountains, an area of hills around a huge crater formed by the impact of a meteorite. NASA wanted to get its hands on some geological samples of lunar sub-soil that had been thrown up by the impact. This mission was ultimately carried out by Apollo 14.

The mission control centre in Houston learns that Apollo 13 'has a problem': one of the service module's oxygen tanks has just exploded. The mission is abandoned, but Apollo 13 has to continue its trajectory, circle the moon and head back to Earth as quickly as possible. On the ground, everyone works out and tests the procedures necessary to bring Lovell, Swigert and Haise home. In this picture you can see the carbon dioxide filter from the command module that has been adapted to fit the LM. Instructions for making it would then be dictated to the stranded crew.

Now living in the LM, the three men circled the moon, seen here, before returning. The command module docked with the LM was switched to standby mode to save oxygen and energy.

As they risked running out of oxygen if they stayed in the command module, the three astronauts took refuge in the lunar module, a temporary but providential life raft, in which they remained until they were back in Earth orbit. But the LM was designed for two people over a fairly short time and the carbon dioxide filtration system was not intended to cope with three men for several days. The command module had plenty of filters, but they were of a different type to those in the LM. One of the filters can be seen behind Swigert, held together with adhesive tape!

that something was wrong. An appraisal of the situation made it clear to the three men that the number 1 tank was damaged and was steadily losing pressure. At this point, no one knew what had caused the explosion. The first thought was that it may have been due to a meteorite impact, but whatever had happened it was obvious that the mission was in trouble and they needed to get back to Earth as quickly as possible. In fact, there had been a short-circuit in one of the tank's cryo stir fans and the resulting spark had created an excess of pressure, causing the tank to explode. Two of the three fuel cells had ceased to function and the craft had begun to change its course slightly. Looking outside, Lovell noticed that something was escaping from the service module. 'It's a sort of gas,' he told Houston. Now, the whole mission needed to be reconsidered. Everyone started to think quickly, both on the ground and in space. The command module's electric power was cut (it needed to be conserved for the return) and the lunar module became a kind of 'life raft' for the crew.

But Apollo 13 had not reached the end of its troubles. Firstly, the crew were not on an immediate-return trajectory – they would have to go around the moon to get themselves onto a return path to Earth. Secondly, the LM, 'Aquarius', had been designed to take two men for two days, not three astronauts for four days; this meant that there was likely to be a build-up of carbon dioxide in the lunar module as there were not enough filters, and those used in the command module were of a different diameter. Because of the reduction in the electricity supply, the ECS (Environmental Control System) had shut down and the inside temperature had dropped noticeably. It was also difficult to transfer the guidance system information from the CSM to the LM, and the crew could not reliably take readings from the stars, as they were frequently obscured by the debris floating outside.

Finally, and more seriously, the SPS engine for directing the craft earthwards could not be used, because its hypergolic propellants were

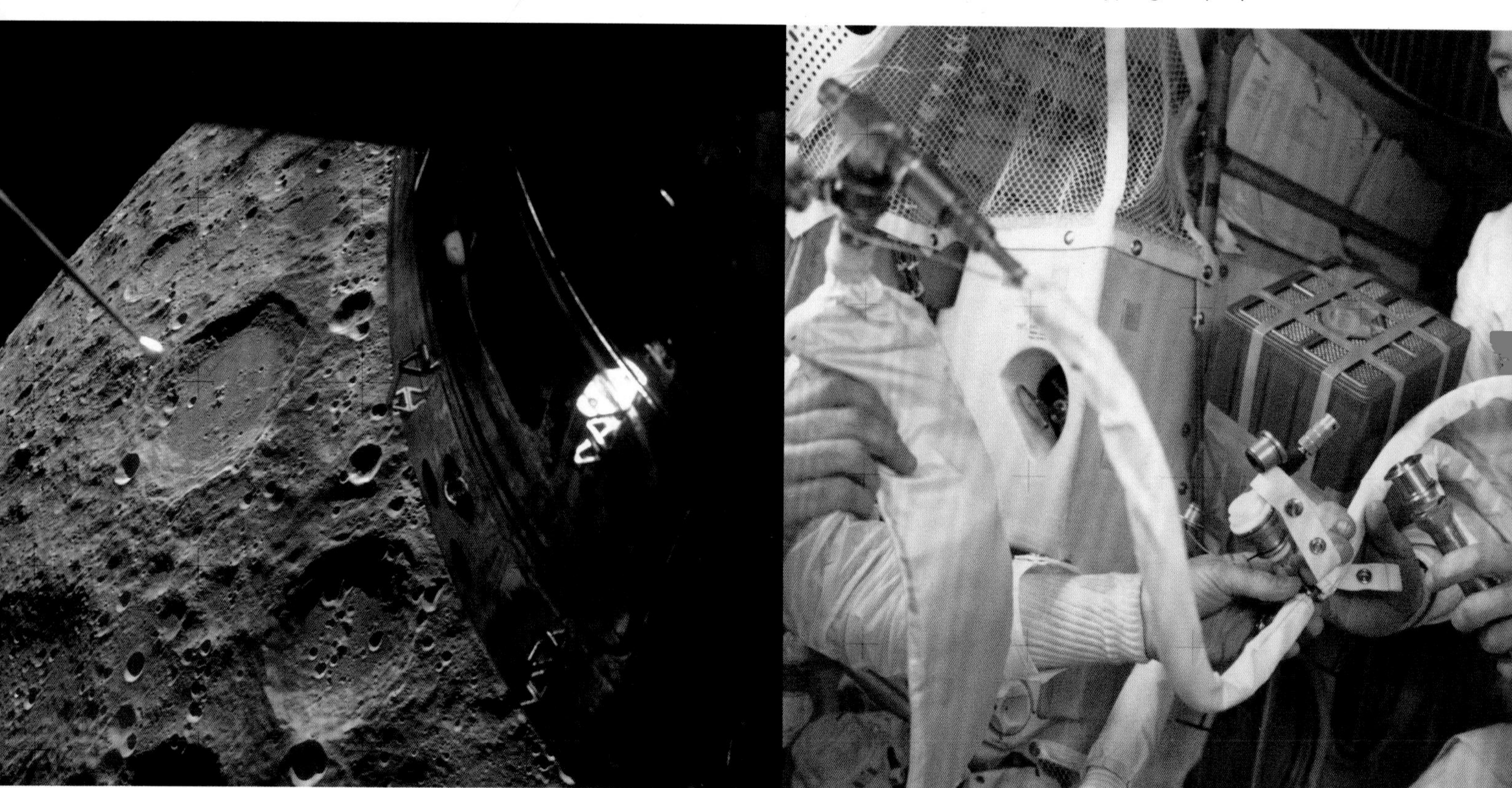

activated by an electrically controlled injection of helium – and, of course, there was no electricity on board! Everything therefore depended on the LM, which had to shelter the crew and bring them back to Earth. Water was also produced by the fuel cells and its principal job was to keep the LM's electronic systems cool. The augury for the return journey was not looking good. To thirst and cold was added the discomfort of the LM. The crew of Apollo 13 were truly 'space-wrecked', and it was only the lunar module, desperately clinging to the CSM, which saved them. Its DPS descent engine was finally used to escape lunar orbit and to make adjustments in the return trajectory.

The CO_2 problem was resolved using plan B. The LM's filters were round, whereas the CM's were square. On the ground, they took stock of what was on board and came up with some do-it-yourself filters made from adhesive tape, a sock and other bits and pieces. It appeared to work, so they sent instructions to the crew, who were able to put together some emergency filters that would purify the ambient air. It worked a treat! The return journey was charged with tension, as they were not sure, after the command module's systems had been shut down for several hours, whether they would restart because of the risk of condensation caused by the intense cold. Fortunately, they all worked without any problems. After returning to the command module, the crew jettisoned the service and lunar modules. From 'Odyssey', Lovell, Swigert and Haise could see the extent of the damage, a section of the cladding having been completely ripped off. Now strapped into their seats, the crew returned safely to Earth. Only Haise was in some discomfort, with a urinary infection. But they had escaped the worst. Indeed, if the accident had occurred after the moon landing, during the return flight, it is unlikely that they would have had enough fuel on board the LM to get back.

The crew of Apollo 13 heaved a sigh of relief when the command module splashed down. Tired, they were at long last back on American soil, in the shape of the amphibious assault ship USS *Iwo Jima*.

The service module has finally been jettisoned and the astronauts have rejoined the command module for their return to Earth. The LM, 'Aquarius', has also been cast off. Despite the mediocre quality of this photograph, the damage is clearly visible. The panels have been blown off by the explosion of the tank.

Stuart Allen Roosa, Alan Shepard and Edgar Dean Mitchell pose for the official photograph for the eighth manned Apollo mission. Originally they were to have been Apollo 13's crew, but not having completed their training they changed places with Lovell, Swigert and Haise. Shepard was the only astronaut from the Mercury programme to serve on Apollo. Roosa, who would remain on the CSM, 'Kittyhawk', would make no further flights after Apollo 14; nor would Mitchell. The latter was very interested in paranormal phenomena and tried an extra-sensory experiment during his flight to the moon, though this was not, of course, authorised by NASA. Although Shepard had been the first American to journey into space, Apollo 14 was only his second space flight. They were soon christened 'the all-rookie crew'.

A view of the control room just before the CSM moved around to dock with the LM.

On each Apollo mission the crew had to install several pieces of scientific equipment on racks around the LM's descent module. Shepard and Mitchell gave the first tests to the Mobile Equipment Transporter (MET) that was used to bring back samples collected some distance from the LM. The mission was much more scientifically based than the two preceding ones. In the background you can see the MET and the CPLEE (Charged Particle Lunar Environment Experiment), designed to study particles from the moon. On the far right, under a film protecting it from the sun's rays, is a device to measure tremors. The small box in the foreground is the battery supplying electricity for all the experiments.

Shepard plays golf!

American public opinion had held its breath, faced with a potentially catastrophic event. The first survivors of a 'space-wreck' had shown that technology had its limits and that man's ingenuity could help him to survive. Apollo 14 struggled to revive the American spirit of conquest, despite the presence of Alan Shepard, one of the original Mercury project astronauts, in command. However, it must be understood that the mission took place in a grave international context. At the start of 1971, the United States was in a very difficult position in Vietnam. In March 1969, the heavy bombing of suspected Viet-cong bases in Cambodia had damaged Nixon's presidency. The withdrawal of troops was beginning, but the American commitment was underlined when the Marines and the USAF supported South Vietnamese troops in the invasion of Laos. In the United States, many people were struggling to understand what was happening. Already bogged down in South-East Asia, were GIs now going to have to fight in Cambodia and Laos against an invisible enemy?

On 31 January 1971, Apollo 14 lifted off with Alan B. Shepard Jr, Stuart A. Roosa and Edgar D. Mitchell on board. For the first time in the history of Apollo flights the countdown had been halted, for 40 minutes, just eight minutes before launch, because of weather conditions. The mission was to be the same as Apollo 13's, but was delayed six months by the enquiry into the Apollo 13 accident, allowing the crew a full 19 months of training. Everything went as planned into Earth orbit, but the docking between the CSM ('Kitty Hawk') and the LM ('Antares') was difficult. They made six attempts at it before Houston came up with a new procedure. By 4 February, the crew were in Earth orbit and the jettisoned Saturn-IVB was heading for a crash-landing on the moon. The LM finally took Shepard and Mitchell down to the moon's surface, but again it showed its capriciousness when, during the preparation for the descent, an alarm went off signalling Houston to abort the mission.

On the ground, there was puzzlement, as nothing appeared to be wrong. The on-board computer seemed to be going wild over nothing, but having received the signal it was about to abort the whole landing procedure. It would have jettisoned the descent engine, activated the ascent engines and initiated docking with the command module without anyone being able to do anything about it! Once again, it was back on Earth that the problem was solved thanks to the engineers at MIT's Instrumentation Laboratory. A new programme was written and entered into the computer in under two hours, thus restoring order.

But the astronauts' troubles were not yet over. At 30,000ft above the moon's surface, the radar regulating the LM's descent rate began to malfunction: it was emitting but not receiving radar waves. With only 10,000ft to go, they would have had to abandon the mission if it failed to work properly. Ground control ordered Shepard to switch the radar on and off, and at 20,000ft altitude, it came back

on again and 'Antares' was able to make a soft landing, 50ft from the intended spot in a mountainous region. The two men would not be short of work, Apollo 14 having a scientific focus with a geological emphasis. Shepard went out first, declaring: 'It took a while, but we've made it.' This comment was a reference to the ten years that had passed between his short hop into space aboard Mercury Freedom 7 and stepping down onto the lunar surface with Apollo 14.

And Shepard had gone through a lot to get there. At one point, because of a hearing problem, he had simply been dropped from the programme. Surgery had allowed him to make a return to service and, despite his 'difficult' reputation, everyone at NASA held him in high regard. Driven but experienced, he was the eldest of the 'moonwalkers'. After a few moments of excitement, Shepard and Mitchell deployed the ALSEP (Apollo Lunar Surface Experiments Package) and made use of the MET (Modular Equipment Transporter), a kind of space-cart – soon nicknamed the 'rickshaw' – which allowed them to transport the containers and equipment needed to gather samples. At the end of the second EVA, Shepard even permitted himself the luxury of a game of golf! He had brought along the iron from a golf club and fixed it to the end of a rod, amusing himself by hitting off several balls. This provoked something of a debate, with some Americans finding it hard to understand quite why so many millions of dollars had been swallowed up just to allow someone the opportunity to play golf on the moon...

In the event, this was a record-breaking mission: the astronauts brought back nearly 100lb of rocks, carried out two EVAs lasting nine hours and 17 minutes, and the LM and its crew had spent 33 hours on the moon. Roosa, left alone in orbit, had been able to carry out numerous experiments and survey the programme's future landing sites. A new, quicker docking procedure with the CSM had been used for the first time and the return flight passed without any problems. During this period, the crew did a number of experiments that were turned to good account with Skylab and later the Space Shuttle, including the study of liquids, electrolysis and hot moulding techniques on various materials. It was also the last time that a lunar crew was kept in quarantine.

Scott, Worden and Irwin, the three Apollo 15 astronauts. After Apollo 9, Scott again took part in the programme with two other new recruits who joined at the same time, on 19 April 1966.

Driving on the moon!

Apollo 15 was almost totally scientific in nature, this type of mission being designated a 'J Mission'. The CSM module was fitted with equipment (the SIM, or Scientific Instrument Module) designed to map the lunar surface, and a small satellite was even put into orbit. Furthermore, this was the first mission to try out the Rover or Lunar Roving Vehicle. This was a sort of 'moon buggy', designed by Boeing and powered by 200-watt electric motors on each wheel. Folded, the Rover was especially compact, occupying a space 5ft by 1.8ft and weighing 460lb when empty. It could carry two astronauts and their equipment and travel at about 7.5mph for 40 miles.

On 26 July 1971, David Scott, Alfred Worden and James Irwin, aboard the command module 'Endeavour', were blasted into space by a Saturn V rocket whose trajectory had been slightly lowered southwards. This allowed insertion into a lower orbit (100 miles), thereby saving fuel while increasing the payload by almost 1,100lb. Instead of the usual eight rocket engines, the launcher had only four, but they were fired for a longer period. As the lunar module was to stay longer on the moon, its batteries and solar panels were larger, hence heavier.

So as to be able to sit easily on the Rover, the crew had received new spacesuits with a revised dorsal pack. Scott and Irwin joined the LM ('Falcon') on 30 July and landed on the moon further on than the intended site, near the Hadley Apennines. Scott went outside the upper airlock for about half an hour, taking

Apollo 15's great novelty was the Lunar Rover Vehicle (LRV), quickly renamed the 'moon buggy'. The rover, made by Boeing, was folded up in a compartment on the LM. It allowed astronauts to go much farther afield in their exploration of the lunar surface. Apollo 15's moonwalkers also wore an improved spacesuit with better articulation around the hips, making it easier to adopt a sitting position on the LRV.

pictures and commenting on the surroundings, but the first proper EVA took place the following day. The proceedings were more or less identical to those of the other missions, with the deployment of the ALSEP, then the setting up of the Rover, stowed in one of the LM's external compartments. This gave the crew a few awkward moments, but the problems were quickly resolved and EVA-1 was able to go ahead with the two men collecting numerous samples of lunar rocks. EVA-2 took place the following day. It was during this excursion that the American flag was erected, and a point three miles from the LM was reached, despite some problems with the steering. The third excursion took the total EVA time to 18 hours and 35 minutes.

The collecting of samples proved to be very interesting, as the astronauts found a piece of rock dating back to the original formation of the moon. This sample, numbered 15415 came to be known as the 'Genesis stone', a very old type of rock that the astronauts had actually been hoping to find. In fact they were unable to complete all their experiments, as some of the rocks were too hard to drill. Scott was very interested in geology, unlike Shepard on the previous mission, and he was able to study and collect samples at his leisure without any interference from Houston. Before finally returning to the lunar module, Scott proved definitively in front of the whole world Galileo's theory on the fall of objects in a vacuum. In front of a movie camera, he dropped a hammer and a feather, which both hit the ground at the same time leading him to conclude that 'Galileo was right…'

Scott and Irwin also set up a small statue in honour of lost astronauts and cosmonauts, before moving the Rover away from the LM with its camera aligned so as to record the take-off. Back in Houston, the geologists and engineers could hardly have been more satisfied: the colour pictures sent back were of superb quality and gave a very clear idea of the moon's appearance. With the crew back together, Worden, the command module pilot, went outside to recover the films from the SIM's bay. 'Endeavour' finally re-entered the Earth's atmosphere on the 13th day. One of the three parachutes did not open fully, but this had no appreciable effect and the capsule came down less than a kilometre from the intended site.

Crews	Duration of flights
Grissom-White-Chaffee	—
Komarov	1 day and 2 hours
—	8 hours 36 minutes
—	11 hours 10 minutes
McDivitt-White	9 hours 57 minutes
Eisele-Schirra-Cunningham	10 days and 20 hours
—	3 days
Beregovoy	3 days and 23 hours
Lovell-Anders-Borman	6 days and 3 hours
Shatalov	2 days 23 hours and 20 minutes
Volynov-Yeliseyev-Khrunov	10 days and 1 hour
McDivitt-Scott-Schweickart	10 days and 1 hour
Stafford-Young-Cernan	8 days
Armstrong-Aldrin-Collins	11 days and 18 minutes, including 59 hours on the moon
Shonin-Kubasov	4 days 22 hours and 42 minutes
Filipchenko-Volkov-Gorbatko	4 days 22 hours and 40 minutes
Shatalov-Yeliseyev	4 days 22 hours and 50 minutes
Conrad-Gordon-Bean	10 days 4 hours and 36 minutes
Lovell-Swigert-Haise	5 days and 23 hours
Nikolayev-Sevastianov	17 days and 17 hours
Shepard-Roosa-Mitchell	9 days
—	175 days (manned for 24 days)
Shatalov-Yeliseyev-Rukavishnikov	1 day 23 hours and 45 minutes
Volkov-Dobrovolsky-Patsayev	23 days and 18 hours
Scott-Worden-Irwin	12 days and 7 hours
Young-Mattingly-Duke	11 days and 2 hours
Cernan-Evans-Schmitt	12 days and 14 hours

Chronology of the Apollo and Soyuz programmes

Mission	Launch date	Comments
Apollo 1	27 January 1967	Rehearsal for flight planned for 21 February. Crew killed.
Soyuz 1	23 April 1967	First Soyuz manned mission. Cosmonaut Komarov killed.
Apollo 4	9 November 1967	Mission AS-501. First launch of Saturn V rocket with the CSM. Unmanned flight.
Apollo 5	22 January 1968	Mission AS-204. First flight of Saturn I-B rocket with the LM. Unmanned flight.
Apollo 6	4 April 1968	Mission AS-502. Saturn V's second flight with CSM.
Apollo 7	11 October 1968	First manned Apollo mission. Launch using Saturn I-B. Long-flight test of CSM.
Soyuz 2	25 October 1968	Second Soyuz orbital mission. Unmanned flight; test docking with Soyuz 3 aborted.
Soyuz 3	26 October 1968	Second manned orbital Soyuz mission; test docking with Soyuz 2 aborted.
Apollo 8	21 December 1968	First manned circumlunar mission.
Soyuz 4	14 January 1969	Third manned orbital Soyuz mission. Docking with Soyuz 5. Yeliseyev and Khrunov carry out an EVA and come on board. The three crew members return.
Soyuz 5	15 January 1969	Fourth manned orbital mission. Yeliseyev and Khrunov leave Soyuz 5 and carry out an EVA. Volynov returns to Earth alone.
Apollo 9	3 March 1969	Manned orbital mission. Test of CSM and LM.
Apollo 10	18 May 1969	Second circumlunar flight. Test of LM's descent, without landing on moon.
Apollo 11	16 July 1969	Third circumlunar flight. Moon landing and EVA.
Soyuz 6	11 October 1969	Fifth manned orbital Soyuz mission. Rendezvous with Soyuz 7 and 8. Mission aborted.
Soyuz 7	12 October 1969	Sixth manned orbital Soyuz mission. Rendezvous with Soyuz 8. Docking with Soyuz 8 not possible.
Soyuz 8	13 October 1969	Seventh manned orbital Soyuz mission. Rendezvous with Soyuz 7. Docking with Soyuz 7 not possible.
Apollo 12	14 November 1969	Fourth circumlunar flight. Second moon landing and two EVAs.
Apollo 13	11 April 1970	Lunar landing cancelled. Tank explodes. Circumlunar flight followed by successful emergency return.
Soyuz 9	1 June 1970	Eighth manned orbital Soyuz mission. First long-duration Soviet flight.
Apollo 14	31 January 1971	Fifth circumlunar flight. Third moon landing and two EVAs.
Salyut 1	19 April 1971	First orbiting station.
Soyuz 10	23 April 1971	Ninth manned orbital Soyuz mission. First attempted docking with Salyut 1 fails.
Soyuz 11	6 June 1971	Tenth manned orbital Soyuz mission. First successful docking with Salyut. Death of Soyuz crew on return.
Apollo 15	26 July 1971	Sixth circumlunar flight. Fourth moon landing and four EVAs. First use of the Lunar Rover.
Apollo 16	16 April 1972	Seventh circumlunar flight. Fifth moon landing and four EVAs. Second use of Lunar Rover.
Apollo 17	11 December 1972	Eighth circumlunar flight. Sixth moon landing and two EVAs.

Eugene A. Cernan driving a lunar rover on the Apollo 17 mission.

The lights go down

The Apollo 16 and 17 missions marked the end of the Apollo era and the moon programme. They took place on 16 April and 7 December 1972 respectively, and were identical to the preceding mission. Both were commanded by experienced men – John W. Young and Eugene Cernan – who were veterans of Gemini and early Apollo. The missions made much use of the Rover, allowing each of them to bring back a hundred kilos of rocks. Apollo 16 had serious problems with the CSM's engine during entry into lunar orbit. While this had little effect on the conduct of the programme, the mission was shortened by a day for safety reasons. Apollo 16 also brought back the largest rock collected from the moon, weighing 24lb.

The Apollo 17 mission was distinguished by being the only night-time launch of the programme and by including a scientist in its crew for the first time, the lunar module's pilot, Harrison Schmitt, being a geologist. To this day, he remains the only 'civilian' to have walked on the moon.

Schmitt took this memorable picture, known as 'the blue marble photo'. It has been reproduced countless times in geography textbooks.

The programme ended on a high note, with Apollo 17 breaking all records and bringing back even more data. When Cernan climbed back into the LM to leave the moon for good, he said: 'I believe history will record that America's challenge of today has forged man's destiny of tomorrow.' The curtain then came down on the Apollo programme. From now on, going into space had to be profitable. NASA had explored, now it had to exploit. Putting aside the purely human achievement, a number of doubts remained, of which the first was: If it's just a matter of taking photographs and bringing back a few pebbles, robotic probes could do that and much more cheaply! The (over-) ambitious programme of Thomas O. Paine, who still defended the idea of going to the moon, brought about his replacement by James C. Fletcher. The Apollo 18 and 19 missions were cancelled, but it was under his mandate that the final three Apollo missions took place, complemented by the success of the Mariner planetary probes. Fletcher maintained NASA's assets while accepting at the same time its reduction in manpower. He brought NASA's administration into the era of pragmatism.

On Apollo 17, three lunar excursions were carried out in the Taurus-Littrow region, which was bordered by high mountains. While the moonwalkers did numerous experiments on the soil, Evans, in the CSM, was not idle: he mapped the moon using some new instruments, providing better information about the state of the ground below the surface.

raised into a higher orbit above the Earth. For a while Mishin considered sending two spacesuit-equipped cosmonauts to make a close examination of the airlock, but this posed a number of problems, starting with the provision of the suits themselves: as we have seen, Soyuz flights were carried out wearing a simple flying suit. Spacesuits were made individually for each cosmonaut, and since Soyuz's crew had no need of them none had been made. Kaminin, as the cosmonauts' leader, then considered the design of the airlock itself and the problems with the automatic docking system, the origins of which dated all the way back to 1962. Several modifications were eventually made to the docking system, and everything went as planned when Soyuz 11 lifted off on 6 June 1971 and docked without any problems. Dobrovolsky, Patsayev and Volkov entered Salyut 1, turned on the lights like someone arriving back home and started up the various systems.

The Soviet crew was to beat the record for time spent in space, two years before the American Skylab station. Eleven days later, the cosmonauts had quite a scare when a serious fire broke out on board. Fortunately they were able to put it out and, on 20 June, they passed the milestone of 1,000 orbits around the Earth. But the men's physical condition was deteriorating. They had a treadmill available, to which they could harness themselves in the weightless conditions, but it was very noisy and shook the whole structure of Salyut, causing the solar panels to wave. Their physical activity was thus curtailed. On 26 June the cosmonauts started their preparations for the return, making a list of the food and drinking water left for the following crew. Finally, on 29 June, they rejoined Soyuz 11 and began the separation manoeuvre from the main station. Patsayev took a number of photographs of Salyut before proceedings for the return got under way.

On the ground, several people were struck by the fact that the crew did not reply to some of their messages, but put this down to poor radio communications, as everything appeared to be going very well on board. The craft was retrieved as usual, but when the recovery team opened the hatch they were horrified to find three lifeless bodies inside. A valve that normally opened at an altitude of $2^1/_2$ miles, equalising the internal and external pressures, had instead opened during the separation of the descent and re-entry modules. The three cosmonauts had died from asphyxiation at an altitude of about 100 miles. The space endurance record had ended in a national tragedy. Without spacesuits, the crew had stood no chance. Future Soyuz flights were reduced to crews of two, both wearing spacesuits.

CHAPTER 4

The space station and Shuttle era

While the Soviets had succeeded in launching the first space station with Salyut 1, this was largely because they had had the time to prepare: out of the race to reach the moon, they were able to concentrate instead on how to live in space. But the Americans were not left behind and, after Apollo, they too were able to develop their Skylab space station, putting to good use some of the 'remnants' of the Apollo programme that had been curtailed for budgetary reasons. In any case, Skylab had always been an integral part of the Apollo programme.

The dream of establishing a permanent presence in space had not begun with either Salyut or Skylab. A variety of projects were developed during the 1950s. When Gemini came to an end, the Air Force wanted to move quickly into the MOL (Manned Orbiting Laboratory) programme based on Gemini-B, developed from McDonnell's Gemini capsule, but attached directly on top of a cylindrical laboratory. The whole thing was to be launched together and, once in orbit, an airlock at the heat-shield end allowed the crew to enter the MOL. The Gemini capsule would then be put on stand-by and reactivated for the astronauts' return to Earth. The MOL was for military purposes and could remain in orbit for 40 days. The programme reached an advanced stage – it had even been approved by President Johnson in 1965 – and was being developed at the same time as Apollo, but by the end of the 1960s it had been abandoned.

Salyut and Skylab provided the opportunity to discover how well men could live permanently in space. More generally, space laboratories would give a chance to study the Earth, to better understand animal and plant

The American Air Force came up with a number of projects for manned space stations, including this one, which spawned a variety of projects, such as the MOL (Manned Orbiting Laboratory). Though well advanced, the project was abandoned in 1969 to save $1.5 billion… Some of the development, however, was transferred to NASA.

Skylab in orbit above mother Earth.

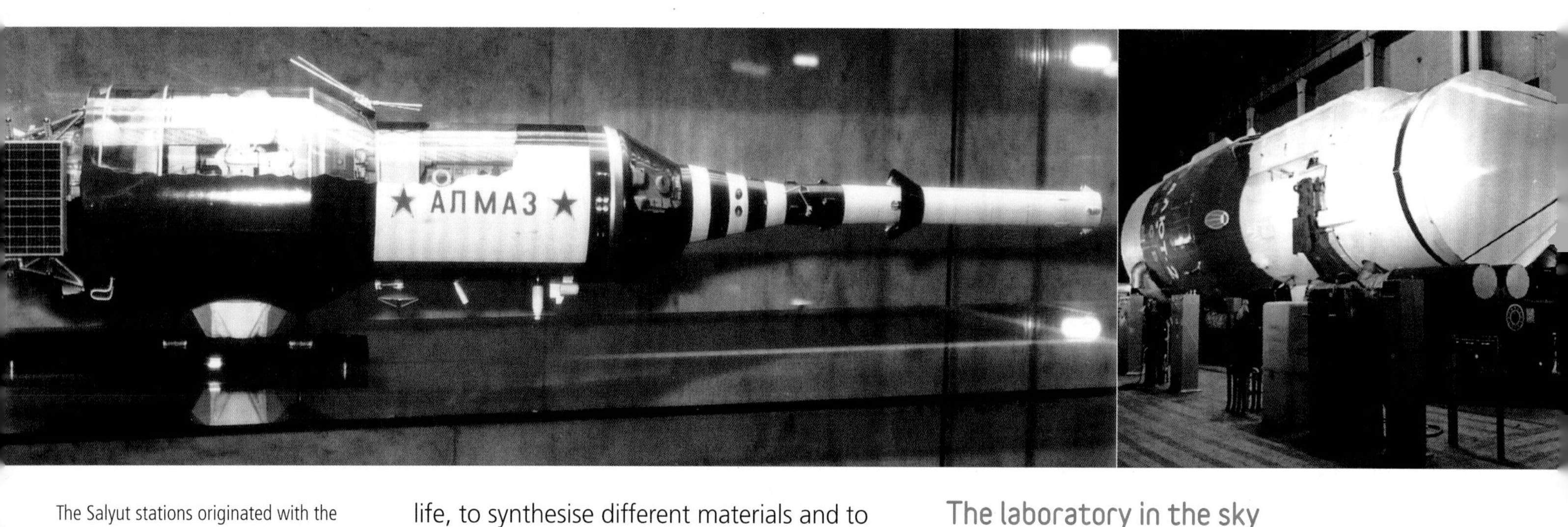

The Salyut stations originated with the Almaz manned military stations characterised by the presence of a large telescope. So as not to draw attention from the West, all the space stations were given the 'Salyut' name. In Soviet terminology, the Salyut 1 civil station was DOS-1 and the military Salyut 2 was in fact the Almaz-1 (or OPS-1). The picture depicts a model of Almaz-3 (OPS-3), alias Salyut 5.

Above right: This is Salyut 2 in its final configuration, ready to be attached to its Proton launcher. It was the first Almaz station, but it was never operational and was soon scrapped. Things were slightly complicated by the fact that the Soviets hid this failure from the West, which did know about the station. So as not to reveal its true nature, they gave it the catch-all designation of 'Cosmos 557'.

life, to synthesise different materials and to organise space rendezvous with a view to building more highly developed stations in the future. After the Soyuz 11 accident, the Soviets decided not to send up any further craft. Not only did Soyuz need modification, but so too did all the procedures and the training of Soviet cosmonauts, one result of which was to switch from three to two cosmonauts per mission because of the extra bulk of their spacesuits. Abandoned to its fate, Salyut 1 burned up on re-entry into the atmosphere on 11 October 1971. There was then a long few months' wait for a new station to be readied. However, the DOS-2 space station did not even survive long enough to receive the Salyut name, as the rocket carrying it blew up on launch in July 1972.

The Russians were so concerned that the Americans might launch Skylab first that Brezhnev had the Almaz project resuscitated. It was thus a military craft, but to deflect the West's attention it was christened Salyut 2. Launched by a Proton rocket at the beginning of April 1973, the station worked normally at first, but after a few days the specialists on the ground noticed that it had depressurised. A crew could not therefore go inside and the planned Soyuz flights were cancelled. It was found that some launch debris had pierced one of the nitrogen tanks providing pressurisation. Unusable, Salyut 2 shut down and after just 54 hours in orbit it was destroyed as it re-entered the atmosphere. After these repeated failures, the Soviets decided to try out the new Soyuz. Soyuz 12 lifted off on 27 September 1973 and stayed only two days in space with Vasily Lazarev and Oleg Makarov on board. In the meantime, the Americans had managed to put Skylab into orbit.

The laboratory in the sky

At the beginning of the 1960s, NASA launched the idea of an orbiting space station to be put into service around the start of the 1970s. The station was to serve not only as a laboratory, but also as a repair and maintenance workshop for spacecraft. Using the powerful Saturn C-5 (Saturn V), it was possible to put this station into orbit in two stages, each of the modules then being put together in space. The initial programme was very ambitious, so that one way or another they would keep ahead of the Russians and Salyut, and preserve American superiority. Over the course of time, the AAP (Apollo Application Programme – the name Skylab would be adopted later) was slimmed down. To reduce costs, the AAP would use a number of pre-existing components. The conquest of the moon was the priority and the Apollo spacecraft and the Saturn launcher would have to be used for this programme. It was for this reason too that the S-IVB final stage of Saturn was to be converted into one of the station's modules. It was in fact the final stage of the Saturn IB AS-212. The idea was to use two Saturn IB launchers to send up the two parts of the AAP. But since the Apollo 18, 19 and 20 missions had not

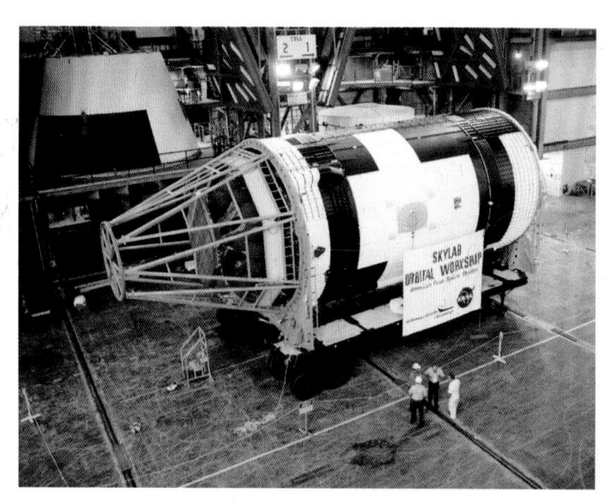

The Skylab laboratory was created from an S-IVB stage from the Saturn IB rocket (the AS-212 rocket that never flew). With its tanks removed, this stage formed the whole of Skylab with its equipment, supplies (stored in containers) and fitting out.

Opposite: On 14 May 1973, Saturn V lifts off on its way to put the station into orbit.

UNITED STATES
USA

A Saturn IB took the first crew up to Skylab. Mission SL-3 consisted of Conrad, Kerwin (a scientist) and Weitz. They confirmed what NASA already knew: Skylab was damaged and would need costly repairs.

taken place, there were three Saturn C-5 (Saturn V) rockets available.

The firm of McDonnell-Douglas was given the job of transforming Saturn IB's final stage (a huge fuel tank) into the OWS (Orbital Work Shop) by adding a docking system. In 1969, after much discussion, von Braun and Gilruth eventually opted for an integrated solution, using a single Saturn V rocket to put Skylab into orbit in its entirety – except for the crew, who would go up to the station in a standard Apollo command module. Naturally, there was much debate at a time when NASA was seeing its annual budget melting away like snow in the sun, but ultimately the extra cost of Skylab was deemed acceptable.

Skylab's strong point was principally its size: with a weight of nearly 100 tons, a length of 80ft and a diameter of over 21ft, it had a usable volume roughly equivalent to that of an apartment, and provided a reasonable degree of comfort. Each of the three crew members who were in residence would have his own living space. Electric power was provided by two wings covered in solar panels, and four other units supplied the astronomical laboratory and its ATM telescope (Apollo Telescope Mount), the linchpin of Skylab, attached to the station. Everything for the crew had been carefully thought out. Thus, dirty clothes were not washed, but thrown away.

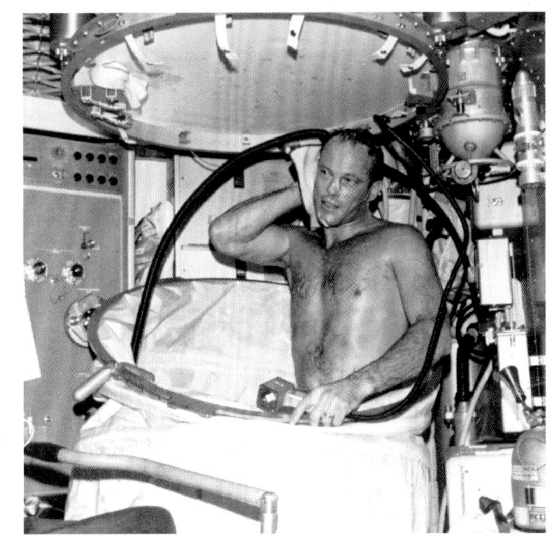

While Garriott works outside, Lousma takes a shower. Life aboard Skylab was distinctly more comfortable than any astronaut had known before.

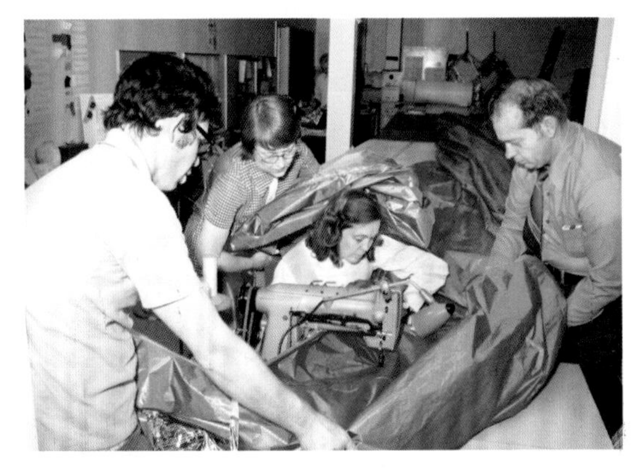

Neatly piled up in cupboards were hundreds of items of underwear, gloves and other clothing, all of which were to be used just once. The luxurious conditions for the astronauts on Skylab included a shower, always difficult to use in the absence of gravity. Liquids would not flow, of course, so they had to be sucked up by a sort of vacuum cleaner to get them to cover the skin. Still, in comparison with other missions the presence of a shower was a welcome innovation.

While looking round, the astronauts from the Skylab 2 command module realised that not only was one of the solar panels missing, but that the second one was wrongly deployed, as can be seen in this close-up photograph. Furthermore, ground control had detected excessively high temperatures inside, caused by the original heat protection having been torn away during lift-off. Hastily, a thermal sheet was cut out and sewn together to be used by the crew.

The final Skylab mission (SL-4), with Gerald Carr, William Pogue and Edward Gibson, was also the longest, the three men staying in Skylab for 89 days. Each of the missions included a scientist chosen by NASA who was not drawn from the military pilots' group. This would eventually become the norm on future shuttle missions. Skylab was an orbital laboratory that conducted a great deal of civilian research.

Skylab was easily identified by its windmill-shaped solar panels that were designed to supply electricity for the telescope. The station remained 'one-armed' for its entire career, its main solar panel never having been replaced. The last mission ended at the beginning of 1974. Skylab remained in orbit until 1979, before being lost through lack of maintenance because the shuttle was not yet ready and there were no further Saturn launchers available.

On 14 May 1973 Skylab was launched from Cape Canaveral's pad number 39A into an approximately 270-mile-high circular orbit. It was then that the ground controllers realised that something was not quite right. Sensors indicated an abnormally high internal temperature and the electricity supply seemed excessively low, yet the ATM's four solar panels had deployed correctly since the first orbit. The problem arose from one of Skylab's two solar wings – which was no longer there… As the Saturn rocket was going up, it had been knocked off by the anti-meteorite shield (also intended to protect the station from solar rays), which had also disappeared. Skylab was therefore alternately positioned with its solar panel facing the sun (to obtain maximum electricity), then moved to cool it off again. Most of the on-board systems were switched off to conserve electricity. This procedure went on for ten days before a Saturn IB brought up Pete Conrad, Paul Weitz and Joseph Kerwin to Skylab. Their main task was to take a look at the extent of the damage.

The first Skylab mission turned out to be somewhat eventful. The astronauts discovered that not only had the anti-meteorite shield and the left solar panel disappeared, but an errant piece of the shield had also caused the second panel to be wrongly deployed. The CSM docked properly, but the crew could not open the airlock. They were thus obliged to repair Skylab using an EVA. The CSM was undocked from Skylab and moved in as closely as possible to the wrongly deployed panel. Weitz became the space mechanic, with Conrad at the controls of the CSM and Kerwin firmly holding on to his legs as the upper part of his body projected from the hatch. Armed with long cutters, he tried unsuccessfully to cut away the offending piece of metal. In the meantime, so as not use up too much fuel, they had to re-dock with the workshop module. Unfortunately, they had to do this no fewer than eight times! By now very tired, the three men preferred to wait until the following day before attempting to enter Skylab.

When they finally boarded they first equipped themselves with breathing apparatus in case the high temperature inside Skylab should have caused gases to form. An EVA was then mounted to attach a sort of parasol to the outside to lower the ambient heat, which reached as high as 52°C. Thanks to this extraordinary deflector – devised by the Houston JSC team in less than a week – the temperature was reduced to under 24°C. Two weeks later, on 7 June, Conrad and Kerwin carried out a second difficult EVA and managed to fully deploy the solar panel.

The crew remained on board Skylab for 28 days and carried out numerous experiments, including a third EVA to change the film canisters. In the event, the mission fulfilled all of its objectives and above all showed that human beings were capable of repairing a damaged spaceship. On 22 June, Conrad, Weitz and Kerwin returned to Earth. Rarely had astronauts taken on so many roles: in less than a month, they had been simultaneously pilots, technicians, repairmen, astronomers and biologists.

Two other fruitful missions were to follow. The first, Skylab 3 (SL-3), took off on 28 July

1973 with Alan Bean, Jack Lousma and Owen Garriott aboard. As intended, the CSM was launched into a lower orbit and docked with the station five orbits later. However, a problem arose with the Apollo module's thrusters. They were able to dock properly but the problem persisted, and engineers on the ground went as far as to prepare a rescue module. (Skylab actually had two docking systems, so it was possible to evacuate a crew whose CSM was defective.) However, as the problem did not put the module's safe return at risk the rescue operation was cancelled. When Bean, Lousma and Garriott went on board, they were surprised to find three other occupants already there – in the shape of hastily-prepared dummies left by the previous crew! Skylab 3 was to break all endurance records at more than 58 days, and its crew accumulated 1,085 hours of work (medical experiments, observation of the Earth and the sun etc). Lousma and Garriott carried out 13 hours and 44 minutes of EVAs.

Skylab 4 and its crew (Gerald Carr, William Pogue and Edward Gibson) took their mission length up to 83 days. They also completed 33 hours and 21 minutes' worth of EVAs, mainly to replace the ATM's film cassettes and carry out other maintenance tasks. Indeed, Skylab's crew initially complained of being constantly on the go: such a lengthy mission required them to make a full inventory, checking the amount of food and drinking water available etc. Nonetheless, the mission gave them the opportunity to observe a number of phenomena for the first time: for instance, thanks to the ATM telescope they were able to follow the comet Kohoutek (discovered only three months before Skylab's launch), and photograph a magnificent solar eruption as well as an eclipse of the same body. Towards the end of the mission, one of Skylab's three gyroscopes started to show signs of wear, but the station remained operational. Finally, on 8 February 1974, Skylab 4 started its protracted return to Earth. The station was to receive no further visits, but in the meantime it was put into a stationary orbit. It had been intended to have the Space Shuttle dock with Skylab and propel it into a higher orbit, but the delays to the shuttle programme condemned the station. It would have needed too much work to prolong its life, and on 11 July 1979 it re-entered the Earth's atmosphere and partially broke up over Australia, its debris killing a cow.

Salyut takes up the torch

After Skylab's return, the Soviets tried to relaunch Salyut. The station had not been much of a success and Salyut 3, launched into orbit on 25 June 1974, was actually an Almaz military station. Considerably less imposing than Skylab at only 18.2 tons and orbiting at an altitude of only 168 miles, Salyut 3 was fitted with two solar panels and a detachable

Starting with the Soyuz 12 mission (up to Soyuz 40) a new vessel was developed for visiting the Salyut stations. Christened 'Ferry', the craft was a revised version of Soyuz 11. It retained the docking system with the internal airlock but no longer had solar panels, as electricity was supplied from batteries. Being lighter, the Ferry was able to carry a greater load when supplying the Salyut stations, but was restricted to a two-day flight. There are virtually no photographs of Salyuts 1 to 3 in orbit. For the want of anything better, we have to rely on the drawings provided by NASA. Salyut 4 (right) was a true civilian orbital station compared with Salyut 3, on the left. Salyut 4 is drawn with a Soyuz Ferry craft.

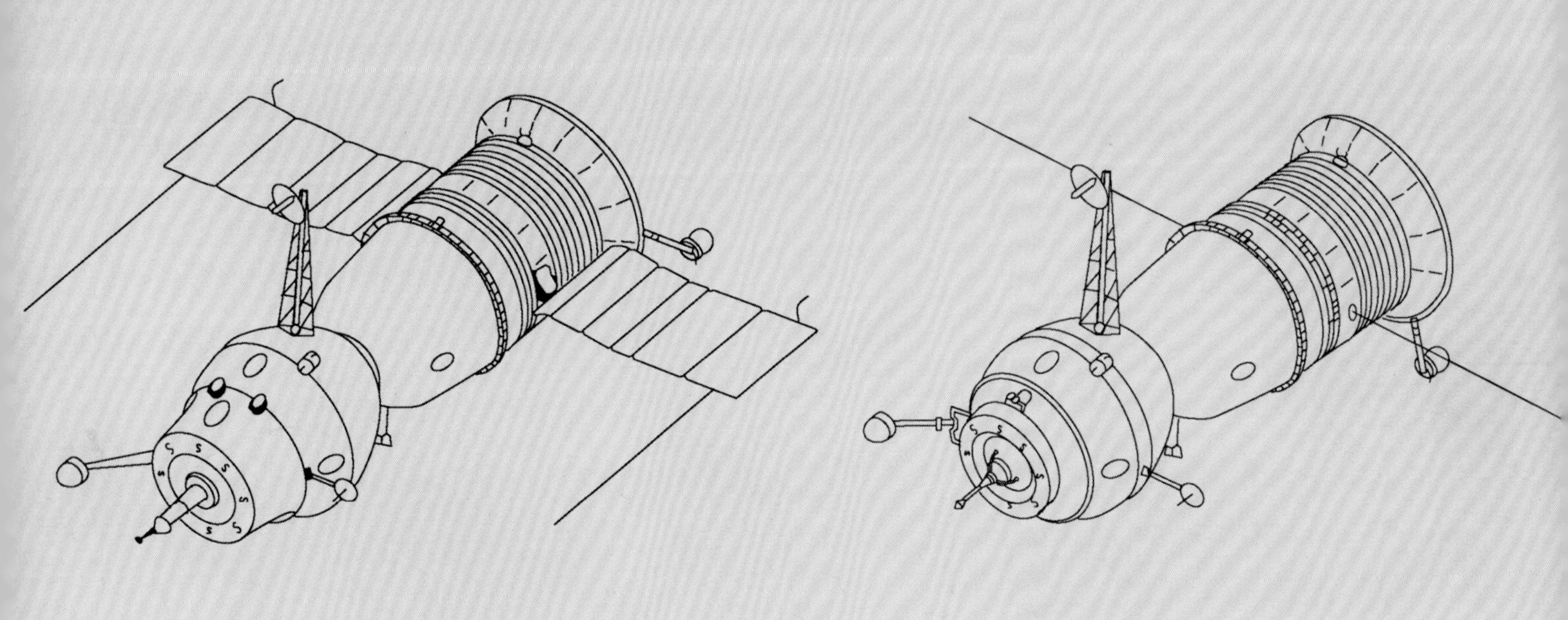

module allowing computer data and films to be sent back automatically to Earth. It was not expected to last more than eight months, but was very well equipped to observe all land and aerial movement on Earth. No fewer than 14 cameras, optical and infra-red sensors (with a 100m resolution) provided the capability for photographing and following a range of targets. Salyut 3 even permitted itself the luxury of locating and taking pictures of Skylab, proving that the Soviets had the capability to track moving objects in space. To maintain continuous contact with the ground, a special ship, the *Yuri Gagarin*, was stationed in the Atlantic.

On 14 July 1974, the Soyuz 14 mission, with Yuri Atiukin and Pavel Popovich, arrived alongside Salyut 3. They stayed for two weeks, undertaking military and medical experiments. Their daily regime of physical exercises proved very beneficial, as when they got back to Earth on 19 July they were able to get out of the capsule unaided. The second mission, Soyuz 15, on 26 August, was less successful. The crew of Lev Demin and Gennady Sarafanov could not manage to dock after the automatic system failed. Being too short of fuel to attempt any further manoeuvres, Soyuz returned to Earth two days later. On 23 September the detachable module began its re-entry into the atmosphere. Although damaged, it brought back its precious data. Finally, in January 1975, an experimental 23mm (or 30mm according to some sources) gun was successfully tested, destroying its target satellite. This put paid to any doubt about Almaz's military purpose. The following day, Salyut 3 was taken out of orbit and burnt up over the Pacific Ocean.

Salyut 4 became operational on 2 December 1974. This time it was a civilian station, the fourth of the DOS type. Originally intended to precede Skylab, the DOS programme encountered numerous problems. Salyut 4 had been modified by getting a third solar panel mounted vertically on top of the main module. Provided with a large 250mm telescope with a focal length of 2.5m, and two tons of equipment, the improved Salyut station was put into an orbit with an apogee of 220 miles. Soyuz 17 joined Salyut 4 on 11 January 1975. Georgi Grechko and Alexei Gubarev appreciated the note left at the station's entrance by its constructors: 'Wipe your feet!' They remained on board for a month, carrying out chiefly astrophysical experiments. The telescope's main mirror had been damaged when it was inadvertently exposed to the full rays of the sun. They repaired it then returned to Earth without any problems.

Soyuz 18 took off on 5 April 1975 with the same crew as Soyuz 12 (Vasily Lazarev and Oleg Makarov). But as the rocket reached an altitude of over 115 miles, the second stage failed to separate fully from the third. The

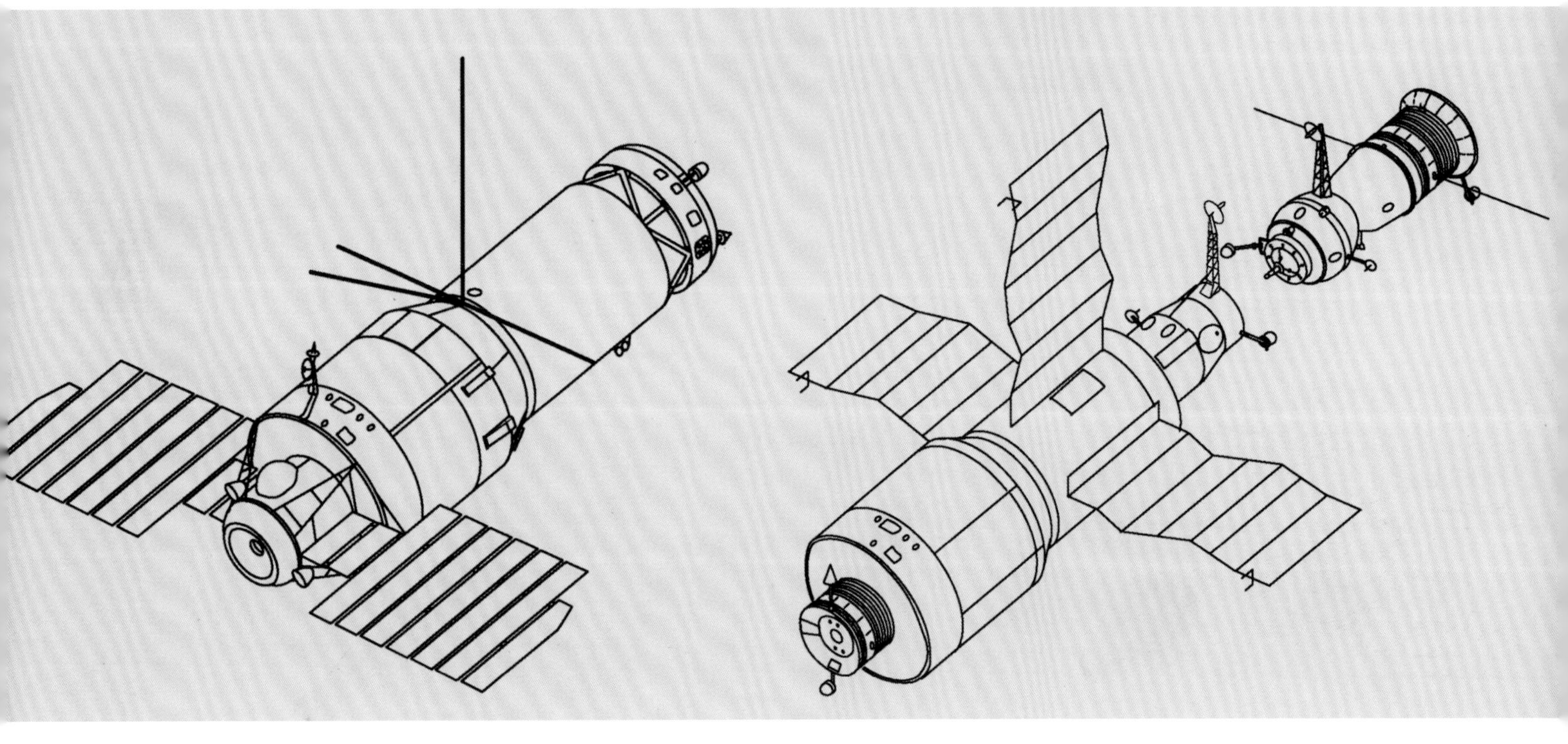

Three Americans and two Soviets are to meet in space. From left to right: Slayton, Stafford, Brand, Leonov and Kubasov. With his two Gemini missions and experience aboard Apollo 10, Stafford is the veteran of this summit meeting, but Slayton is the eldest and the longest-serving with NASA, his selection dating back to the Mercury programme. Grounded for many years, he was resigned to choosing astronauts for Gemini and Apollo. Then in 1973, after an exhaustive physical preparation, he chose himself to take part in the Apollo-Soyuz mission! He would later direct the shuttle's approach-flight programme.

The ASTP meeting was not a spur-of-the-moment event. It had been a long time in preparation; the crews had got to know each other and worked together for weeks. The key piece of hardware in the mission was the adaptor that permitted docking between Soyuz and the CSM. The two countries' engineers had naturally worked together on its design. It was 10.33ft long, had a diameter of 4.59ft and weighed a little more than two tons. The composition of the atmosphere in the two vessels was different (100 per cent oxygen in the American CSM and an oxygen-nitrogen mixture in the Soviet Soyuz). They could not, therefore, mix and the adaptor served as a 'decompression chamber'.

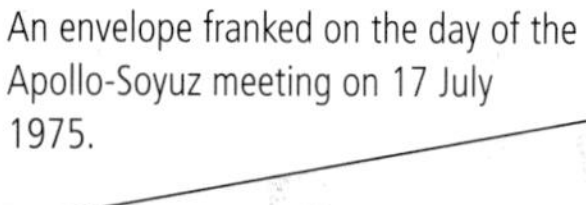

An envelope franked on the day of the Apollo-Soyuz meeting on 17 July 1975.

latter's engines then fired and blew away whatever was still attached. Soyuz 18 began to veer seriously off course, to the extent that the automatic emergency mode was activated, cutting the craft free from the third stage to bring it back to Earth. But as the craft was at that moment pointing downwards it simply accelerated even further, and instead of undergoing 15G as normal the crew was briefly subjected to more than 21G! However, the emergency procedure functioned perfectly: the parachutes opened and Soyuz landed in the middle of the desert near the Soviet-Chinese border. In fact the crew had landed 50 miles inside China, obliging a Red Army helicopter to cross the border illegally to recover the cosmonauts.

The Soviet leaders imposed a complete news blackout on this mission, which was consequently named 'Soyuz 18a'. As far as Brezhnev was concerned, it had never happened. And when the next mission proved a success, it was, logically enough, named Soyuz 18. This took place between 24 May and 26 July 1975 with cosmonauts Vitali Sevastianov and Piotr Klimuk, who remained aboard Salyut 4 for 63 days. They were followed by an automatically-guided Soyuz 20, sent up to check whether a Soyuz coupled to Salyut could remain long enough in orbit before re-entering. Both these missions were a success. The station was taken out of orbit on 3 February 1977 and burnt up in the upper layers of the atmosphere.

There had been a number of changes at the heart of the Soviet research department. The bitter failures of the N1 rocket were at the root of the upheaval and led to a complete reorganisation of the Russian bureau. Mishin was sacked from OKB-1, and Korolev's and Glushko's offices were merged into a single research department named 'NPO Energia'.

The CSM and its adaptor seen from Soyuz. The American CSM was much easier to manoeuvre than Soyuz. The latter was of a new type, specially designed for this mission and fitted with an APAS system (Androgynous Peripheral Assembly System) for docking with the adaptor. The system was called 'androgynous' because it had both male and female connections. So, either Soyuz or the CSM could make the docking manoeuvre. If one of them did not succeed, then the other would make an attempt and the mission would not have to be aborted.

The Saturn IB – like the CSM – was surplus from the Apollo missions, and consequently they did not cost very much. In any event, they would otherwise have ended their days in a museum. With this mission, launches carried out by Saturn came to an end.

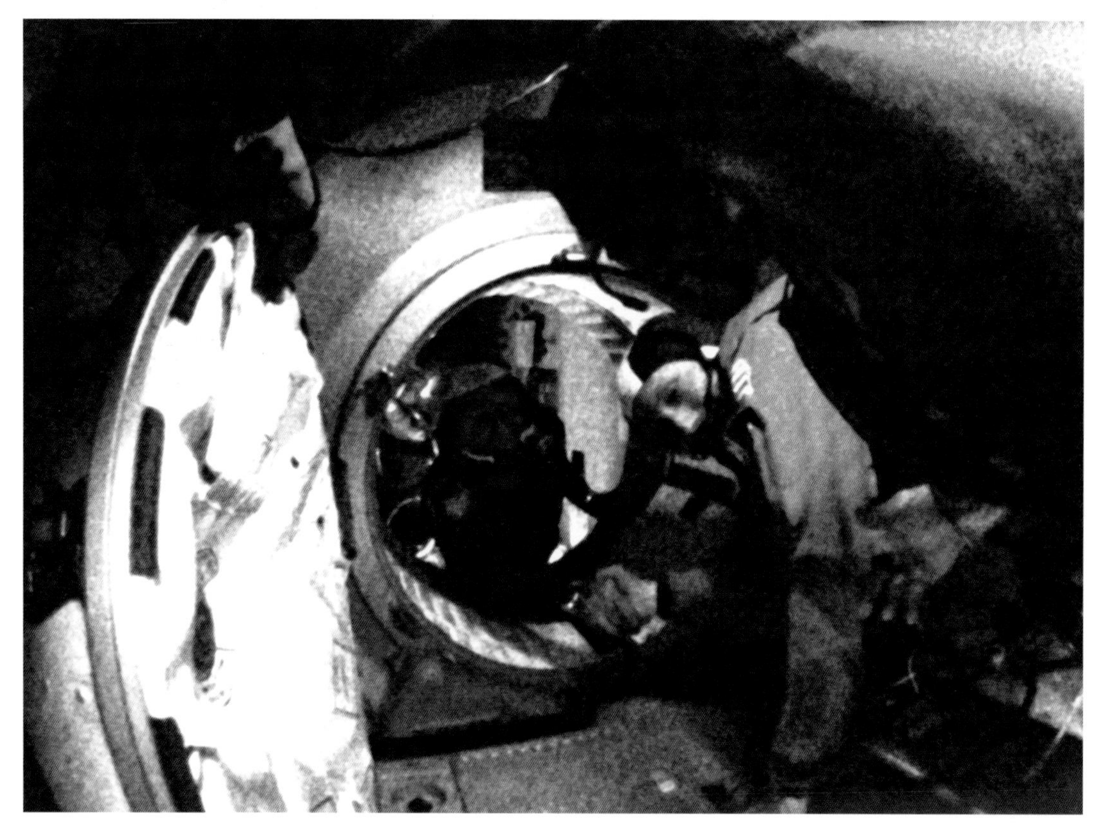

The famous historic handshake. The mission was a complete media success. However, everything had been rehearsed time and time again. Nonetheless, this common desire to work together in space would bear fruit a few years later.

The Apollo-Soyuz interlude

The rendezvous between Apollo and Soyuz 19 remains in the annals of the conquest of space as a summit meeting between two 'enemy' powers. Politically speaking, in 1975 Soviet-American relations were in a period of détente. In May 1972 Nixon and Brezhnev had signed an anti-ballistic missile limitation treaty and negotiations on strategic arms limitation were well under way. Nixon had resigned in consequence of the Watergate affair, but the new US President, Jimmy Carter, was equally in favour of talks with Brezhnev. The ASTP (Apollo-Soyuz Test Project) itself went back to the beginning of the 1970s. At first the two countries worked out an ambitious plan based around Skylab meeting up with a Salyut station. This was then revised to a simple emergency common docking system, so that a craft from either country could rescue the other's crew.

The project was given the go-ahead in 1972 and planned for 1975. A Soyuz craft was modified and its new systems tested by the Soyuz 16 mission, with Filipchenko and Rukavishnikov on board, in December 1974. In order to prepare for the future ASTP flight, Soyuz 16 was even followed by the Johnson Space Center. This gave rise to some argument, with the Soviets not being keen to publish details of the flight before the launch, whereas NASA, with its usual desire for openness, wanted them to be known. Eventually a compromise was reached and the flight went perfectly. On 15 July 1975, Soyuz 19 took off from Baikonur carrying Alexei Leonov and Valeri Kubasov, while the Apollo ASTP with Thomas Stafford, Vance Brand and Deke Slayton lifted off from Florida seven and a half hours later. The Saturn IB rocket was carrying the indispensable docking module that would form the 'bridge' between the two spacecraft.

By 17 July Apollo and Soyuz were in sight of

The Soyuz 19 ASTP craft has been specially modified so that it can dock with the Apollo module. It has an androgynous docking system – clearly visible here with its jaws – and a descent module for two men. The green colour had been chosen in agreement with NASA to make it easier to see the craft.

one another and they docked without any trouble. Three hours later, the two crews had made contact. An historic handshake took place between Stafford and Leonov, followed by the exchange of gifts. For 44 hours the crew moved from one module to the other. Several undocking and re-docking manoeuvres were practised and they worked together on various experiments – for example, Apollo shone a laser beam onto a reflector carried by Soyuz 19, then positioned itself in front of the sun to create an artificial eclipse so that Soyuz's crew could take pictures of the sun's corona. Once the two craft had finally been separated the docking module was jettisoned and each crew then continued with its own experiments. Apollo remained in orbit until 24 July, with Soyuz having already returned on the 21st. This brought the Apollo programme to a definitive end. Technically speaking, the ASTP mission was hardly complex, and for NASA it was by no means costly, as they were able to make use of a launcher and a module that had been 'stockpiled' after the end of the moon programme. Nevertheless, it was a well-managed publicity and political event.

To test the modified Soyuz craft for the ASTP mission and to reassure NASA, the Soviets sent up Soyuz 16 with Filipchenko and Rukavishnikov. Everything went as planned.

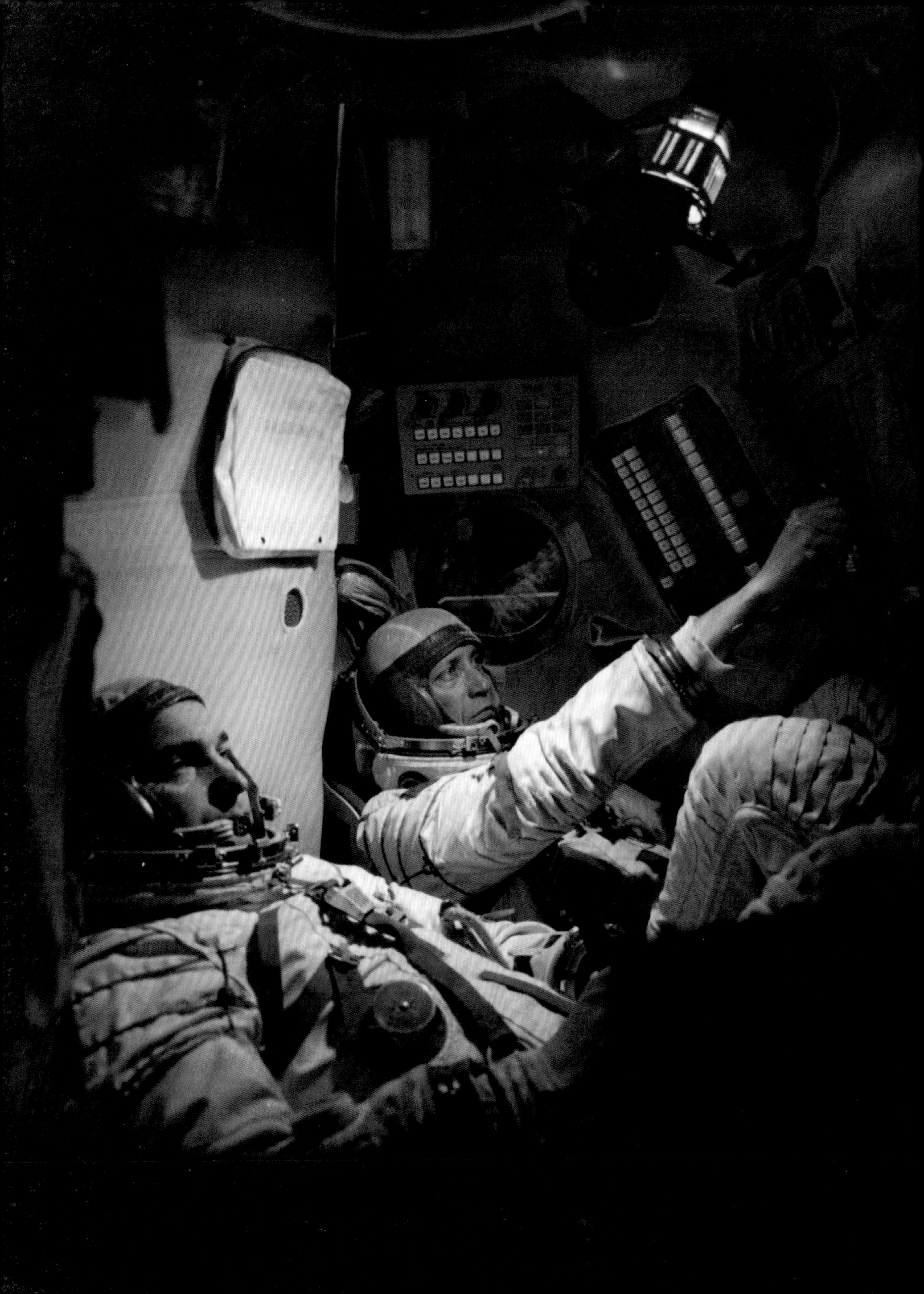

Opposite: Soyuz 22 (with cosmonauts Bykovsky and Aksionov) was the Soyuz replacement used in the ASTP link-up. It retained almost the same equipment, but the 'androgynous' docking system had been replaced with a camera. Soyuz 22 was used for a simple Earth-observation mission, in particular of East Germany. Note the semi-reclining position of the crew, with bent knees, and the very cramped nature of the module.

Volynov and Zholobov were the first to join Salyut 5. Although a large part of their work was observation of the Earth for military purposes, they developed certain procedures that would later be used in civilian stations. Among these was fuel transfer between vessels, which would be used with the Progress craft. After nearly 50 days in orbit the two men made a rather hasty return to Earth. The reason is not known, but it appears that they had suffered from some fairly severe psychological problems.

The Salyut story continues

After 1975, manned American flights ceased. NASA concentrated on sending up probes and on the huge space plane project. But the Soviets had not finished with their Salyut programme, which was to become the driving force in the Russian space programme. It was thanks to this programme that the Mir stations were developed, and the experience gained would also be a determining factor in their design for the great ISS (International Space Station) project. Despite the rapprochement with the United States, illustrated by the Apollo-Soyuz meeting, the Soviets sent a new Almaz military station, Salyut 5, into orbit on 22 June 1976. It was the last ship of this type and it remained in operation for 412 days. Similar to Salyut 3, but with two docking ports, Salyut 5 was to be visited by two crews between June 1976 and February 1977 (Soyuz 21 and Soyuz 24), being occupied for a total of 67 days. Soyuz 21's crew had to leave the station in a hurry for reasons that are unclear. (They may have been victims of smoke poisoning following a leak of fumes.) Soyuz 23, launched on 14 October 1976, was unable to dock properly with the station because of an electronic fault. Furthermore, its crew, consisting of Rozdestvenski and Zudov, almost froze to death while waiting for the recovery teams after landing in a frozen lake with a water temperature of -20°C!

The leap forward occurred with Salyuts 6 and 7, put into orbit on 29 September 1977 and 19 April 1982 respectively. While the external appearance of these two space dwellings was not dissimilar from the other DOS-type stations, these Salyuts were substantially modernised and much more ambitious. Firstly, there were two docking systems, one at each end. One of them was designed to take the new Progress craft, a supply module capable of bringing up 5,000lb of fuel, food, equipment and spare parts. Progress was, in fact, just a Soyuz craft stripped of the equipment needed for

Salyut 6 was part of the second generation of space stations. More modern, but retaining the classic shape of the older Salyuts, this station had two docking points. Salyut 6 had six crews making long-term visits, over a total of 683 days. It was also the first station to receive a cosmonaut who was neither Soviet nor American, in the person of the Czech Vladimir Remek.

The crew of Soyuz 26, consisting of Yuri Romanenko and Georgi Grechko, was the first to man Salyut 6, arriving at the station on 10 December 1976. They had to make an EVA to inspect the docking point after the failure of the previous mission. They also tested the new Orlan spacesuit, which has been in service ever since. During the EVA Romanenko forgot to attach his lifeline, but fortunately for him Grechko noticed and grabbed hold of it to prevent him from drifting away.

cosmonauts, and working on batteries (*ie* there were no solar-panel wings). Once it had docked it was unloaded like a delivery van, and, better still, it then became a refuse lorry: rubbish was stored inside until it was full, when it was undocked and sent to its destruction in the atmosphere. Progress was simple to use and remains in service with the ISS.

The second improvement was in the new Salyut station's comfort, with, for example, a proper shower, which the earlier Salyuts lacked. With Salyuts 6 and 7 the Soviets had taken a considerable step forward in space travel. Between them, they amassed a total of 1,499 days of human occupation, being host to 49 cosmonauts – some of whom were from other countries – who came and went at a frenetic rate. Dockings with Soyuz, Soyuz T ('Transport'), Progress and even a TKS became almost routine. During the early Salyut missions the Soviets had not carried out any EVAs, yet Salyut 6 had three to its credit and Salyut 7 no fewer than 13. Man could now spend long periods living and working in space. The other side of the coin was that these missions excited little public interest, even though they were becoming internationalised.

Salyut 6 was the first station to receive two Soyuz vehicles, 26 and 27, the latter linking up on 10 January 1978. The engineers on the ground were a little concerned about potential structural weaknesses in the station, but everything went well and both two-man crews worked together for a number of hours. In April 1980 Soyuz 36 docked with the Salyut 6 station. Leonid Popov and Valery Rumin would then spend 185 days aboard a station just 47.2ft long. During their stay, the six members of the crew who followed each other welcomed ten 'visiting' cosmonauts. On Salyut 7, Leonid Kizim, Vladimir Soloviev and Oleg Atkov took the record to 237 days during 1984. Salyut 7 hosted the first Frenchman in space, Jean-Loup Chrétien spending eight days aboard in June and July 1982. Svetlana Savitskaya, the second woman in space (and the first to do an EVA), was welcomed aboard on 11 July 1984. In all, Salyut 7 was occupied by six more-or-less permanent crews who were visited by four others. Salyuts 6 and 7 were almost identical and, when the first of them burnt up on its re-entry into the atmosphere after 1,764 days in orbit, the second had already been in orbit for just under three months. This had only been possible because Salyut 7 was effectively Salyut 6's double.

Designed as the backup for Salyut 6 in case it was lost, Salyut 7 ended the Salyut era with a flourish. The station remained 3,216 days in orbit, in other words a little under nine years, a remarkable achievement. It was fitted with solar panels whose size could be increased by adding secondary panels to boost the electricity capacity. Note the Soyuz craft docked at the lower end.

The Space Shuttle: a true false revolution

The Space Shuttle concept took on concrete form during Nixon's presidency. From 1969, after the men of Apollo 11 had set foot on the moon, the American president engaged in a policy of cost reduction. Apollo had swallowed up close to $135 billion dollars, of which $46 billion was for the Saturn launchers alone! Bearing in mind that each one of them was lost after launch, it was vital to pursue the idea of a reusable space vehicle. After the moon, orbiting space laboratories and manned space stations seemed logical developments. These stations would allow observation of the Earth, and would serve as bases for future space exploration. A supply ship shuttling back and forth – the name was readymade – seemed to be the path to follow, as not only would such a cargo ship have lower initial costs, but it would also be cheaper to run. Indeed, it could effectively replace all the existing launchers. Being reusable, it would be more productive, reducing the per-kilo costs of putting a payload into space. Naturally enough, this idea appealed to a Congress that had to vote on NASA's budgets, especially as the country was going through a period of galloping inflation.

Experiments conducted by the engineers at the Marshall Space Center had shown that the

The forerunners of the Space Shuttle had diverse origins and purposes, but they shared a common aim: to create a recoverable space vehicle that could re-enter the atmosphere either gliding or flying. A standard rocket has serious drawbacks: it takes off vertically, consumes a lot of fuel, is expensive, and can be used only once. Fitting rockets with wings was considered, having them take off on a slightly inclined trajectory so as to benefit from the wings' lift, but this proved to be an uneconomic solution. Indeed, the distance to be travelled through the atmospheric layer is greater than with a vertical lift-off, thus requiring more thrust, a higher power output and greater expense.

By combining a standard rocket and a space plane, the engineers found a halfway-house solution: the Dyna Soar project (or X-20), which was to be launched on a Titan rocket and would then return to base on its own. This project was not very far removed from that of the Space Shuttle. But the X-20's wings were of no use during take-off and were something of a dead weight. For this reason NASA developed the 'Lifting Bodies' project, using craft without wings that were able to glide thanks to their shape alone and make a piloted return to Earth.

The 'Lifting Bodies' family of planes had a direct influence on the shuttle design, which was, however, fitted with a delta wing at the request of the Air Force.

first stage of a Saturn IB – the most expensive part, with its five Rocketdyne F-1 engines – could remain immersed in seawater for several hours and be put back into operation after heavy repair. Boeing, the designer of this stage, had put forward an interesting proposition: it could be fitted with thermal protection, stabilisers equipped with air brakes and four big parachutes that would slow its descent to the sea. This system was not retained, through lack of time, but it was not completely forgotten as the method of recovery of the shuttle's two boosters was to be based on this principle. It was always the first few kilometres of a flight that used the most fuel. To escape the Earth's gravitational pull and achieve the necessary speed required tons of thrust. The speed would gradually increase, but less fuel was needed for acceleration.

The role of the X-15 – half-plane, half-rocket – was also a determining factor. It was able to provide information about sub-orbital flight and atmosphere re-entry techniques.

The idea of recovering a part of the launcher so that it could be reused had been running through the heads of NASA's engineers for some time. The two massive SRBs (Solid Rocket Boosters) that launch the Shuttle are thus recoverable. They generate 70 per cent of the power needed to help the orbiter escape the Earth's pull. Here one of the first SRBs is seen under static test at the facility of its builder, Thiokol.

Reusable craft become the priority

From a technical point of view, vertical launch of a spacecraft seemed the quickest way to pass through the dense layers of the atmosphere. A standard take off could not be contemplated, as it lengthened the trajectory substantially and required a large amount of fuel. One other appealing idea was the combination of a booster engine and a space engine. The booster would take off using a kind of catapult, taking the orbiting craft to an altitude of around 20 miles. The latter would then separate and, under its own power, go into orbit at an altitude of about 180 miles. These two piloted, recoverable aerofoil craft would have run on hydrogen and liquid oxygen. The project – called RT-8 – was the brainchild of Eugene Sanders of the German firm Junkers. He had also come up with an ambitious programme for the USAF, the Air Force Dynamic Soaring or 'Dyna Soar'. This winged craft glided back to Earth and was the object of some very costly study lasting from 1958 until 1962.

For their part, NASA's Flight Research Center engineers designed a series of 'lifting bodies' from 1962 onwards. These experimental craft were intended to test the concept of a reusable space plane at a time when parachute-suspended capsules coming down in the sea were the only method of returning a crew safely to Earth. The study of the aeronautical phenomenon of lift was at the centre of the development of these odd prototypes, even though some of them flew more like smoothing irons, but thanks to their remarkable design they were able to land at a much steeper approach angle. There is no doubt that these flights yielded very valuable information and the final form of the Space Shuttle owes much to such strange birds. The North-American X-15, capable of sub-orbital flight as early as 1959, was also of inestimable value. Then in January 1968 Max Hunter, an engineer at Lockheed, put forward the idea of a 'stage and a half'. This was an independent spacecraft attached to enormous tanks, which, once empty, were jettisoned and then recovered.

All these ideas remained in a state of perpetual change until it finally became necessary to make a decision, when, at the end of the 1960s, the United States set its sights firmly on the construction of a very large space station. However, everything almost came to a premature end when the space station project foundered.

The shuttle gets the go-ahead

Nixon's administration hoped to scrap the Space Shuttle programme as being of no value, but NASA turned to the Department of Defense. The shuttle would effectively meet military requirements, as it could put much heavier payloads into orbit than could the USAF's Titan rocket. The Air Force was keen, but imposed its own conditions – a minimum payload of 18 tons, a delta wing, a range of over 1,200 miles – and then gave it a name: the ILRV (Integrated Launch and Re-entry Vehicle). The presidential U-turn came in 1971. James Fletcher – who took the reins at NASA in April 1971 – found an unexpected ally in the person of Caspar Weinberger, the assistant manager of Management and Budget, a federal body charged with keeping control of government spending. By this time the pre-projects for the shuttle had been started and several manufacturers, such as General Dynamic, Lockheed, McDonnell-Douglas, Martin-Marietta, North-American Rockwell and Pratt & Whitney Rocketdyne, were already working on the concept; but nothing had yet been approved by the ultimate power, namely President Nixon.

Eventually, under pressure from Weinberger, Nixon gave the green light. Now named the Space Transportation System, the shuttle went onto the drawing board in 1972, after various rethinks that delayed its entry into service by at least two years (1981 instead of 1979). Despite the decision to go ahead with the shuttle, NASA had undergone drastic cuts: after the abandonment of the Apollo programme and its final two missions, the second Skylab went into a museum and only four out of the envisaged five shuttles were authorised. The shuttle therefore had to be particularly productive to justify its development costs. This would have disastrous consequences in the long term, as, in order to increase the number of launches, maintenance of the shuttles was neglected, to the extent that after the accidents to *Challenger* in 1986 and *Columbia* in 2006 the Air Force withdrew from the programme, considering it too dangerous.

In the summer of 1972, North-American Rockwell (which now belongs to Boeing), the engine manufacturer Rocketdyne, and Martin Marietta (now in the Lockheed-Martin group) received orders to construct the shuttle's body (also known as the 'orbiter'), engines and disposable fuel tank respectively. Thiokol Propulsion – currently Alliant Tech Systems – was to make the two boosters. In 1974, work was started on the construction of STS *Enterprise*, a demonstration shuttle that would not go into space but provided the opportunity to test the craft's behaviour during the return flight phase. For this, the shuttle was carried on the back of a Boeing 747 SCA (Shuttle Carrier Aircraft) bought from American Airlines and modified by NASA.

In its definitive form, the shuttle was to be a reusable spacecraft, a sort of 'space-truck', fitted with a movable arm to grab satellites, to place them into orbit or indeed to bring them back down in its hold. On take off, the shuttle comprised three elements. The boosters (solid-fuel rockets) were mounted on either side of a huge hydrogen and liquid oxygen fuel tank, on top of which was the shuttle itself. Three of the latter's engines were fed by the tank and were used for lift-off, but the main thrust (83 per cent) came from the boosters. These were jettisoned after about two minutes at an altitude of around

The SSME (Space Shuttle Main Engine) is one of the three main engines used in lift-off. Developed by Rocketdyne, these engines are supplied with oxygen and liquid hydrogen from the huge external tank (ET). Here, one of the engines is being lifted for a test in complex A-2 of the John C. Stennis Space Centre.

The orbiter's imposing external tank being wheeled out of the Michoud shops on 9 September 1977. This first tank was only ever used for static tests. It is painted white, but after a few missions this was no longer done, thereby saving around 600lb of weight. Several ET variants were built, all with the aim of reducing the weight. From mission STS-6 onwards, the LWT (Lightweight Tank) was used. Everything used to give the tank rigidity was either modified to save weight, or lightened by the use of new materials (titanium for the booster attachments). Mission STS-91 in 1998 was the first to use the SLWT (Super-Lightweight Tank), which principally made use of an aluminium and lithium alloy. Every kilo shed allowed the payload to be increased by the same amount, increasing the orbiter's productivity. However, the extra cost of the SLWT compared with the LWT is substantial, and the latter is consequently still a part of NASA's inventory.

Fred Haise, one of the crew members of Apollo 13, took part in five pre-flights on the Shuttle *Enterprise*. These were the well-known gliding approach and landing flights (ALTs), lasting just a few minutes after launch from the Boeing 747 carrier plane. During this programme, Haise's fellow crewmember was Gordon Fullerton, who made an orbital flight on the Shuttle with mission STS-3.

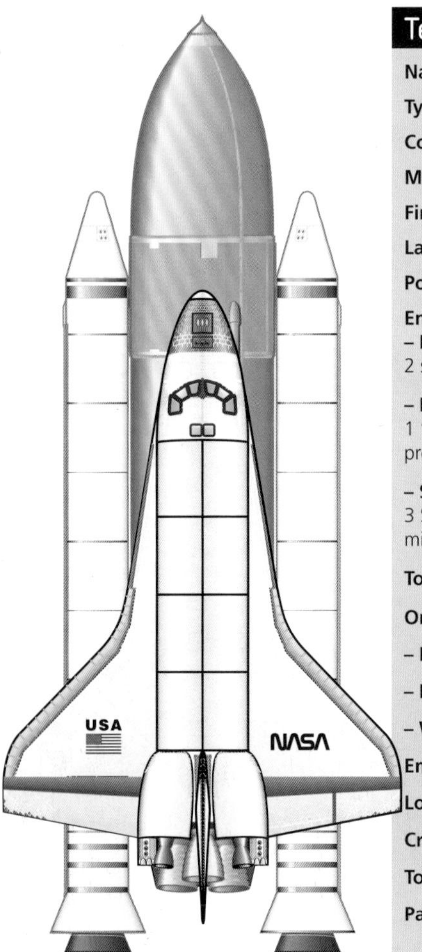

Technical details

Name: *Space Shuttle*

Type: STS Space Transportation System

Country: United States

Manufacturer: North American

First launch: 12 April 1981

Last launch: —

Power on lift-off: 25,751.60kN

Engines:
– First stage (0):
2 solid-fuel SRB rockets of 11,520kN thrust for 124 seconds

– First stage (1):
1 'ET' tank containing liquid oxygen and liquid hydrogen providing thrust for 480 seconds

– Second stage (2):
3 SSME engines running on an oxygen/liquid hydrogen mixture giving 6,834.30kN of thrust for 480 seconds

Total weight on lift-off: 1,997.56 tons

Orbiter:

– Length: 122.17ft

– Diameter: 16.08ft

– Wingspan: 76.35ft

Empty weight: 67.5 tons

Loaded weight: 107.2 tons

Crew: from 2 to 10 (normally 7)

Total height: 56.67ft

Payload: 27.41 tons

30 miles. The second element was the ET (external tank), which was jettisoned after $8\frac{1}{2}$ minutes at about 70 miles and was destroyed on re-entering the atmosphere. The first two missions carried white-painted external tanks, but as the paint increased the weight by 600lb the engineers subsequently decided to leave the tank the colour of its brown insulating material. To protect the orbiter on re-entry, thousands of thermal tiles were affixed to the places that would get hottest. The bulk of the orbiter's construction was in aluminium, but the engine section was made from titanium. For the first time, use was made of a 'fly-by-wire' avionics system where all flight control was achieved not by mechanical linkages via hydraulic circuits, but by computer-controlled electronic commands. The shuttle was designed for about 100 launches, or an average lifespan of around ten years.

The Shuttle *Enterprise* never went into space, being used to make test flights in the atmosphere. Without engines and thermal tiles, it could not have flown anyway. To improve the aerodynamics, a cone covered the rear of the Shuttle. *Enterprise* was to have been named *Constitution*, but the name was changed as a result of a competition organised by NASA (around 200,000 letters were received), and fans of the TV series *Star Trek* thus got their way!

The Boeing 747 SCA (Shuttle Carrier Aircraft) was used to launch the Shuttle during the ALT flights, and also to transport the Shuttles between landing zones and launch pad. There were in fact two SCAs: a standard 747 and a short-range version. The first was a former American Airlines jumbo jet modified in 1976, while the second was bought from Japan Airlines in 1988. Flying with a Shuttle on its back considerably reduced the Boeing 747's performance and its range was a fifth of normal. It was therefore impossible to complete a transcontinental flight in one go and stops had to be made about every 900 miles.

The first flights in the atmosphere; the first orbital flight

The first shuttle was used purely for ALT (Approach and Landing Test) flights to test its aerodynamic characteristics in the atmosphere. *Enterprise* was officially completed on 17 September 1976 and taken by road to Edwards Air Force Base on 31 January 1977. In February, it was mounted on the back of the Boeing SCA 747. From that date a number of manned and unmanned flights took place in this fashion, followed by the final stage of the ALT programme concentrating on free flight. Launched from the Boeing 747, *Enterprise* made only five test flights, instead of the planned eight. These lightning flights – between release and landing at Edwards took only two to five and a half minutes, depending on the kind of release – allowed the landing parameters from a gliding approach to be established. Indeed, during descent, the engineless shuttle resembles a huge glider with a ridiculous lift-to-drag ratio of just three! The 'lift-to-drag ratio' is the relation, at a given

Columbia – which took its name from the first American ship to circumnavigate the Earth – made the first orbital flight on 12 April 1981 with John Young and Robert Crippen. One of the distinctive features of this Shuttle was the presence, clearly visible here, of the American flag on the left wing and the letters USA on the right wing. These were not present on the other Shuttles. During this first flight, with a reduced, two-man crew, the ejector seats were operational. *Columbia* broke up while re-entering the atmosphere on 1 February 2003, killing its crew.

speed, between the lift and the aerodynamic drag, without the assistance of an engine, or, put more simply, the ratio between the horizontal distance travelled and the altitude lost. A racing glider usually has a lift-to-drag ratio of around 60, which means that it would cover 60 horizontal kilometres for every 1,000m of altitude lost. A modern airliner has a lift-to-drag ratio of somewhere between 20 and 25 and a fighter plane a ratio of 10. With its ratio of 3, the shuttle flies about as well as a stone or a falling helicopter! However, this unique craft goes from a speed of over 17,500mph to about 220mph on landing, which explains its poor ratio, as it would otherwise take hours to come down (with consequent overheating).

The first orbital flight by a shuttle was undertaken by *Columbia* on 12 April 1981. *Columbia* had been built at more or less the same time as *Enterprise* in 1975, and it was to have been operational in 1979 for a hypothetical rendezvous with Skylab. But the Space Shuttle programme was running at least two years late and Skylab was destroyed before the orbiter entered service. Standing on launch pad number 39A at the Kennedy Space Center, the shuttle unquestionably represented the acme of man's technological achievement at the beginning of the 1980s. It easily eclipsed the Saturn-Apollo duo that had stirred American hearts ten years earlier. If further proof was needed, the 3,800 journalists who came to cover the event outnumbered even those who had gathered for the launch of Apollo 11! Of course, this was the first time since Apollo-Soyuz that NASA had been back into space.

Flight STS-1 was entrusted to two mission-hardened astronauts, John Young and Bob Crippen. The first had already gained his wings on Gemini 3 and Gemini 6, and then taken part in the Apollo 10 and Apollo 16 missions. By 1981 he was the most experienced man at NASA. For his part, Crippen had joined the astronaut group in 1969. He had had a fairly quiet career, but had carried out a number of important simulations for the Skylab and Apollo-Soyuz programmes. He would later be called upon to fly on other shuttles.

Planned for 10 April, the STS-1 flight was postponed because of a fault in the main computer. The countdown proceeded normally. NASA wanted no problems, with this being the first manned flight for some years, the first manned mission to be sent into orbit

USA

using solid-fuel boosters, and the shuttle's maiden space flight. At T minus 11 seconds, thousands of cubic metres of water were sprayed over the area, partly to cool it, but also to dampen the vibrations that might damage the shuttle. At T minus 3.8 seconds, the main engines fired. In one second, more than four tons of hydrogen/liquid oxygen mixture were consumed by the three engines, then functioning at 90 per cent of their capacity. At T plus 2.88 seconds, the boosters entered the picture. The burning gases were blasting the launch pad at about 6,000mph. A huge cloud of steam billowed up and began to spread. Then the shuttle, hanging on to the back of its fuel tank, rose slowly into the sky on its two solid-fuel rockets.

Compared with Saturn V, acceleration seemed lightning quick. After two minutes and 12 seconds, the boosters slowly separated and *Columbia* continued its rise at some 3,000mph. After eight minutes and 32 seconds the main engines fell silent, and 19 seconds later the external tank was jettisoned. Another 12 minutes and *Columbia* would be in orbit travelling at 17,500mph. Strapped into their ejector seats – which would be abandoned in later shuttles – Young and Crippen began to breathe more freely. The flight was intended to test the orbiter's flight equipment and review all the systems. The cargo-bay doors were opened twice. In its hold, the shuttle carried a DFI (Development Flight Instrumentation) monitoring system.

Overall, the flight was a complete success. The only concern arose from the shock waves created by the boosters on take-off, which had dislodged 60 of the heat-resistant tiles, while 148 others had been badly damaged. The return phase began after 36 orbits, with the shuttle at an altitude of over 180 miles. It began its descent, then started a broad right-hand turn. At an altitude of about 25 miles *Columbia* started to dip its nose and glide, followed by a sharp 180° turn to the left so as to begin its final approach at an angle of 20°. Lined up on Edwards's runway 23, the orbiter made a perfect flared landing. Nineteen seconds before the final touchdown, the undercarriage was lowered and locked in position. At 180 knots, it landed without any problems.

The test missions continued with STS-2, STS-3 and STS-4, which naturally relied on *Columbia*, being the only shuttle then

On 12 November 1981, *Columbia* takes off on its second, two-day flight with Joseph H. Engle and Richard H. Truly aboard. This photograph taken from an observation aircraft shows the long trail left by the Shuttle and its boosters. Also visible is the slight shift in trajectory made shortly after lift-off.

available. Mission STS-2 was the first mission where a spacecraft had been reused. The launch was postponed until 12 November 1981 (instead of 9 October) because of a number of faults with *Columbia*. It took off with Joseph H. Engle as mission commander and Richard H. Truly as pilot. In its cargo hold, the shuttle had a mobile arm known as an RMS (Remote Manipulator System). A problem linked to one of the fuel cells obliged *Columbia* to return earlier than planned but the mission was nevertheless a success, especially as this time only about a dozen tiles had been damaged.

On 22 March 1982 *Columbia* took off with Jack R. Lousma and Charles G. Fullerton aboard. On this occasion the orbiter remained in space for more than eight days, a day longer than intended because of strong wind conditions at the landing zone, which was not at Edwards but at White Sands in New Mexico. The landing here showed that the shuttle could land on a narrow desert strip that had not been designed for this type of craft.

Finally, mission STS-4 took place on 27 June 1982. Thomas K. Mattingly and Henry W. Hartsfield stayed in the shuttle for over seven days before returning safely to Edwards. With the experimental stage now complete, the shuttle was about to enter its commercial phase.

On the tiles...

The space Shuttle had to undergo a difficult re-entry into the atmosphere. Unlike other craft, it had to be recoverable and then overhauled ready for the next mission. Furthermore, in comparison with the Apollo module it presented a much greater surface area. The means chosen to dissipate the heat and protect the structure was to attach black tiles under the fuselage, around the nose and on the leading edges of the wings – where the temperature reached 1,650°C. These vulnerable points were also reinforced with carbon fibre. The areas less subject to heat damage were covered in white tiles resistant to temperatures of 600° to 700°C. In all, no fewer than 24,000 of these ceramic tiles – actually made from a sort of glass-coated silica fibre – were carefully applied. Rectangles of approximately 6in by 4in and variable thickness, the tiles had to be replaced after each flight. Their weakness lay in the application, which was not always very secure, and in their fragility – they could be crushed in the palm of one's hand. During mission STS-1, several tiles came off on launch and the damage was visible to the crew while in orbit.

Opposite: *Columbia*'s lift-off was watched on television by millions of people. It was, in fact, the first American space flight for nearly six years (the last had been the Apollo-Soyuz link-up). The launch was to have taken place on 10 April 1981, but was postponed to the 12th because of a problem with the on-board computer. At seven o' clock precisely (local time), Florida's blue skies are marked by a huge cloud of steam: the SSME engines ignite and consume their ration of 16 tons of oxygen and hydrogen mixture per second. Three seconds later, the boosters roar into life. The ground trembles, the burning gases spew out at over 6,000mph and around 3.5 million tons of water a minute are sprayed over the launch pad in an effort to keep it cool.

Columbia glides back to Earth almost noiselessly, its 90 tons hurtling down in a flared landing that reduces its speed. At an altitude of 180,000ft, the orbiter is still travelling at Mach 10.3. The speed drops to Mach 6 at 120,000ft, then to Mach 1.8 at 65,000ft. The Shuttle slows by carrying out tight turns with the nose up to offer maximum resistance. The main undercarriage touches down at about 250mph. At just under 220mph, the nose-wheel hits the ground. Everything depends on the orbiter's weight and the wind conditions. After the *Challenger* disaster, a braking parachute was fitted to the Shuttles.

Eating in space

In a weightless state, food is not much different from what we eat on Earth. It is the preparation and packaging of it that changes. All food products taken aboard a spacecraft are pre-packed, as the smallest particles could scatter everywhere and cause potential damage to the on-board electronics. Any food that might create crumbs (bread, biscuits and so on) is thus banned. Preference is given instead to foods having a consistency that 'sticks' readily to eating utensils, such as purées and compotes. It is possible to put salt and pepper on food, but they have to be dissolved in water or oil first.

In the early days of space conquest, on the Mercury and Vostok missions, the flights were quite short. Experts were not sure whether mastication would be effective, allowing food to pass easily down to the stomach, and food was therefore provided either in tubes or as chewable cubes. Such provisions had a high energy content but were tasteless. For the Gemini missions, however, the dishes were more varied in both quantity and quality. The cubes remained, though they were covered in gelatine to prevent small pieces dispersing, but the tubes were abandoned when it was realised that they weighed more than their contents! Dehydrated food in packets or boxes began to appear. These had the advantage of being lighter and easily packaged. As the Gemini capsules had a water supply produced by the fuel cells, it was possible to rehydrate the food using a kind of water pistol to squirt water directly into the packet. Each astronaut chose his own menu before take-off (the options included prawns, eggs and macaroni). The taste was better and each dish provided between 2,500 and 2,800 calories per day.

The Apollo programme resulted in even more improvements, thanks to the presence of hot water: coffee and the dehydrated dishes were now much more enjoyable. (In the Gemini capsules water was never at more than 21°C, as against 67°C on Apollo.) It was now possible to eat using a special spoon.

On Skylab and Salyut there was considerably more space. A 'dining room' was installed and proper meals appeared. Much more diner-friendly, the meals could be heated, and the use of dehydrated food diminished in favour of pre-prepared heat-treated dishes (to destroy bacteria), packed in quick-opening cans like those found everywhere on Earth these days. Tuna, salmon, ham, fruits and puddings could be kept like this for a long time. At the same time, water and fruit juice was provided in crushable packs and bottles, the packaging thus taking up less space when thrown away. The crews ate with standard spoons, knives and forks, attached to a magnet so that they would not float away after use. No fewer than 72 different dishes were brought up to Skylab.

The Shuttle did not have a refrigerator so there was a return to dehydrated, semi-dehydrated and heat-treated dishes; but it was now possible to make up individual meals: tortilla, sandwiches, eggs, mixed dishes and so on. In the era of international co-operation, the variety of dishes multiplied: the Americans discovered Russian food and vice versa. The arrival of astronauts of other nationalities brought about a further increase in variety.

The Progress vessels regularly resupply the space station crews, bringing more fresh dishes, and with the missions being so long the energy requirement has risen to 3,000 calories a day. More recently, a new form of irradiated food has been developed: all enzyme content is eliminated from the dish using ionising rays. This is used principally with fruit and vegetables, as well as lean meat (like beef steak and smoked turkey, both popular with Americans), as other foods rich in lipids take on a sickening flavour when so treated. Alongside the American beef steaks, a French firm has established a good reputation with its pâté de campagne, Alsatian jugged hare and lobster à l'armoricaine! Nowadays, there is even talk of 'space gastronomy'.

Chronology of the Salyut and Skylab programmes and related missions

Mission	Launch date	Comments
Salyut 2 (DOS-1)	4 April 1973	Almaz-type military station. Never occupied. Destroyed 28 May 1973.
Skylab	14 May 1973	Launch of the American orbital space station.
Skylab 2 (SL-2)	25 May 1973	First manned Skylab mission. Launched by CSM's Saturn IB. Three EVAs carried out.
Skylab 3 (SL-3)	28 July 1973	Second manned Skylab mission. Launched by CSM's Saturn IB. Three EVAs carried out.
Skylab 4 (SL-4)	16 November 1973	Third manned Skylab mission. Launched by CSM's Saturn IB. Four EVAs carried out.
Soyuz 13	18 December 1973	New Soyuz craft's second mission. Scientific experiments.
Salyut 3 (DOS-3)	25 June 1974	Almaz-type military station. Never occupied.
Soyuz 14	3 July 1974	Flight to Salyut 3. Military and scientific mission.
Soyuz 15	26 August 1974	Flight to Salyut 3. Not possible to dock with station.
Soyuz 16	2 December 1974	Preparatory test flight for future joint Apollo-Soyuz flight.
Salyut 4 (DOS-4)	26 December 1974	Launch of Salyut 4 orbital station.
Soyuz 17	11 January 1975	Flight to Salyut 4. Scientific mission.
Soyuz 18a	5 April 1975	Aborted flight to Salyut 4. Emergency return to Earth because of incorrect separation of rocket stage
Soyuz 18	24 May 1975	Flight to Salyut 4. Scientific mission.
Soyuz 19	15 July 1975	Apollo-Soyuz ASTP mission.
Apollo	15 July 1975	Apollo-Soyuz ASTP mission.
Soyuz 20	17 November 1975	Flight to Salyut 4. Unmanned scientific mission.
Salyut 5 (DOS-5)	22 June 1976	Launch of Salyut 5 orbital space station.
Soyuz 21	6 July 1976	Flight to Salyut 5. Crew return to Earth prematurely after physical and mental problems.
Soyuz 22	15 September 1976	Manned orbital scientific mission.
Soyuz 23	14 October 1976	Flight to Salyut 5. Mission abandoned after failed docking.
Soyuz 24	7 February 1977	Flight to Salyut 5. Scientific mission.
ALT-12 *Enterprise*	12 August 1977	First free flight of the Space Shuttle Enterprise.
ALT-13 *Enterprise*	13 September 1977	Second free flight of the Space Shuttle Enterprise.
ALT-14 *Enterprise*	23 September 1977	Third free flight of the Space Shuttle Enterprise.
Salyut 6 (DOS-6)	29 September 1977	Launch of Salyut 6 orbital space station.
ALT-15 *Enterprise*	12 October 1977	Fourth free flight of the Space Shuttle Enterprise.
ALT-16 *Enterprise*	26 October 1977	Fifth free flight of the Space Shuttle Enterprise.
Soyuz 25	10 December 1977	Supply flight to Salyut 6. Docking failed.
Flights to Salyut 6		
Salyut 7 (DOS-7)	19 April 1982	Launch of Salyut 7 orbital space station.
Flights to Salyut 7		
-		
The first Shuttle flights		
STS-1 *Columbia*	12 April 1981	Shuttle's first orbital validation flight.
STS-2 *Columbia*	12 November 1981	Shuttle's second orbital validation flight.
STS-3 *Columbia*	22 March 1982	Shuttle's third orbital validation flight.
STS-4 *Columbia*	27 June 1982	Shuttle's fourth orbital validation flight.

Crew	Duration of flight
—	54 days
—	2,249 days (manned for 171 days)
Conrad-Weitz-Kerwin	28 days
Bean-Lousma-Garriott	59 days and 11 hours
Carr-Pogue-Gibson	84 days and 1 hour
Lebedev-Klimuk	7 days and 21 hours
—	213 days (manned for 15 days)
Artiukhin-Popovich	15 days 17 hours and 30 minutes
Denim-Sarafanov	2 days
Filipchenko-Rukavishnikov	5 days 22 hours and 23 minutes
—	770 days (manned for 92 days)
Grechko-Gubarev	29 days and 13 hours
Lazarev-Makarov	21 minutes
Klimuk-Sevastianov	62 days and 23 hours
Leonov-Kubasov	5 days 22 hours and 30 minutes
Stafford-Brand-Slayton	9 days 1 hour and 28 minutes
—	90 days 11 hours and 46 minutes
—	412 days (manned for 67 days)
Volynov-Zholobov	49 days and 6 hours
Bykovsky-Aksionov	7 days and 22 hours
Zudov-Rozdestvenski	2 days
Gorbatko-Glazkov	17 days 17 hours and 26 minutes
Haise-Fullerton	5 minutes 21 seconds
Engle-Truly	5 minutes 28 seconds
Haise-Fullerton	5 minutes 34 seconds
—	1,764 days (manned for 683 days)
Engle-Truly	2 minutes 34 seconds
Haise-Fullerton	2 minutes 01 second
Kovayonok-Ryumin	2 days
—	3,216 days (816 days manned)
Young-Crippen	2 days and 6 hours
Engle-Truly	2 days and 6 hours
Lousma-Fullerton	8 days
Mattingly-Hartsfield	7 days and 1 hour

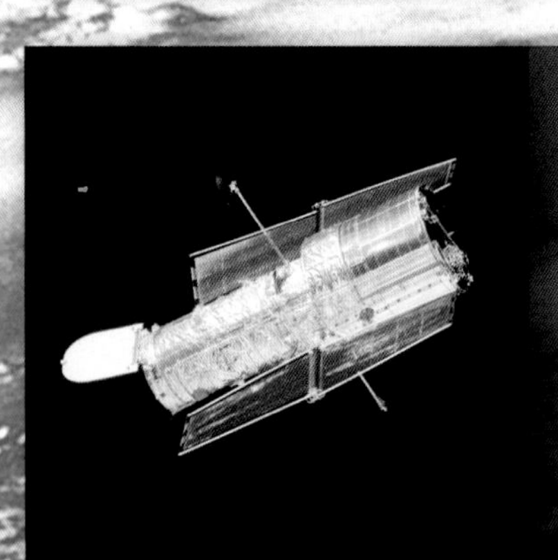

CHAPTER 5

Atlantis

A shared space

Modern manned flight is not so far removed in concept and methods from the manned flights of the 1970s. It is still about sending human beings into space to observe the Earth and carry out various scientific experiments. Because of the cost, exploration of other planets has been confined to probes and, to this day, the furthest a human has ever been sent from the Earth is approximately 248,600 miles. Nobody has yet returned to the moon. NASA and its ex-Soviet counterpart Energia have focused their activities instead on two large orbiting stations. The first is the Russian Mir station, which followed the Salyut programme, and the second is the ISS, the result of international co-operation. While the United States and Russia may remain the unchallenged leaders in space flight, they are no longer the only ones in the field. Initially various foreign astronauts were 'invited' to take part by one or the other, but since 2003 China has also got off the ground by sending up taikonaut Yang Liwei on board the spacecraft *Shenzhou*. This third power looks set to steal a march on Europe and Japan and time will tell whether the arrival of this newcomer in space will change anything.

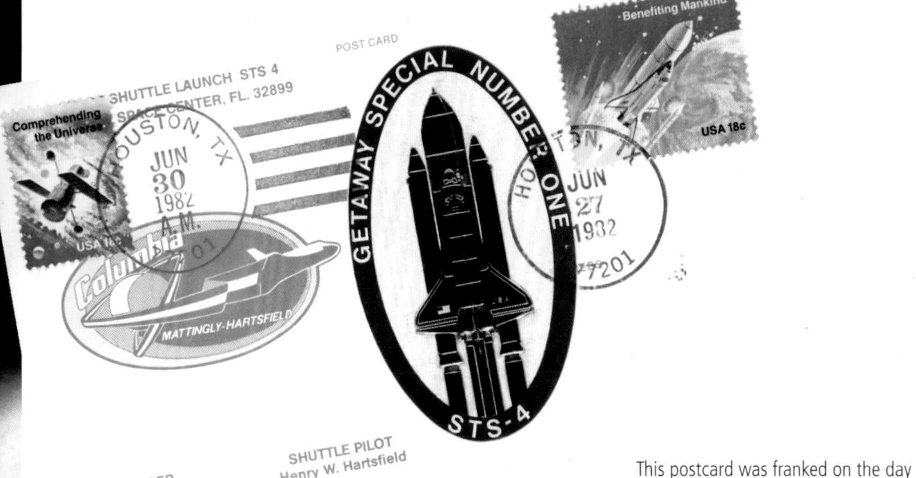

This postcard was franked on the day of the launch of STS-4, *Columbia*'s fourth mission. This final test flight before the Shuttle's entry into operational service was manned by Thomas K. Mattingly and Henry W. Hartsfield. The mission was the first to carry the GAS systems (small containers holding experiments from universities and research centres). After take-off, the boosters – which are normally recovered – were lost because of a defective main parachute.

Sally Ride, the first American woman in space, took part in *Columbia*'s mission STS-7. She had joined the first female astronaut group in 1978, and then acted as Capcom (Capsule Communicator) for missions STS-2 and STS-3. She took part in a second flight (STS-41G) and was then selected for a third orbital mission before the loss of *Challenger* changed the schedule. She later played a major role in NASA, joining the commission of enquiry into the *Challenger* accident, and collaborating on the future role of the United States in space. Since 1989 she has been a Professor of Physics at the University of San Diego, while also directing the California Space Institute. She also set up the Sally Ride Science company, to encourage an interest in science, maths and technology among school children.

The Shuttle fails to make a profit!

The Shuttle's first commercial mission got under way with STS-5 on 11 November 1982. This time there were four men on board, two of whom were specialists, Joseph P. Allen and William B. Lenoir, flown by mission commander Vance D. Brand and pilot Robert F. Overmyer. A payload of some 14.3 tons was taken on, consisting of two communications satellites and a small experimental laboratory of West German origin. This was one of the features of the Shuttle. Since its conception, the idea had been to reserve a part of its hold for small experiment modules in hermetically sealed containers (GAS). They were intended for use by research, educational and foreign or domestic bodies, whether government or otherwise. The plan was to profit from the space in the hold, which was already making money from the presence of a satellite or two, by adding a few extra experiments. In this way costs might be significantly reduced. The two satellites were launched during the STS-5 mission and entered their geosynchronous orbits under the power of their own engines. A planned spacewalk was cancelled because of a problem with the spacesuits. The return took place without pressurised suits on Tuesday 16 November.

Challenger is captured here by a camera mounted on the SPAS (Shuttle Pallet Satellite), made by the German firm of Messerschmidt-Bolkow-Blohm. It was an experimental module that could 'work' from the Shuttle's hold, but could also be put into orbit via the Shuttle's mobile arm, thus becoming an independent satellite. The SPAS-01 was able to carry out a number of experiments, including studies on the behaviour of various types of metal in conditions of minimal gravity. Note also, at the rear of the Shuttle's hold, the white containers that had held two satellites to be put into orbit.

In April 1983 the second orbiting Shuttle was ready to be launched. Christened *Challenger*, it differed principally from its predecessor in that heat protection was partially entrusted to woven Nomex in place of tiles on the upper surfaces. It was lighter and its payload was consequently increased. It was with *Challenger*, on 4 April 1983, that a long series of EVAs began using the new spacesuit specially designed for the orbiter. The TDRS (Tracking and Data Relay Satellite) it carried was launched, but its two-stage engine cut out abruptly, sending it into an elliptical orbit. However, it was able to regain its intended orbit several months later.

Sally K. Ride was the first American woman in space. She joined her four other companions on *Challenger* for the STS-7 mission, which carried a new cargo of two satellites, an OSTA-2 pallet – sensors fixed to a pallet in the hold – and a German SPAS-1 (Shuttle Pallet Satellite) laboratory, which could either function independently from the Shuttle's hold or be put into orbit by the robot arm, thus becoming an artificial satellite. STS-8 was the first mission in which both the launch and landing took place at night. On board *Challenger* was the first African-American in the history of the conquest of space, in the person of Dr Guion Stewart Bluford.

The first real commercial mission for the Shuttle *Columbia* was to put into orbit two communications satellites made by Hughes. The satellites fired their engines 45 minutes after having been released from the Shuttle's hold to put them into a higher, elliptical orbit. Other experiments were also undertaken, including one from West Germany. Mission STS-5 was also the first mission to have more than three people on board. An EVA had been scheduled, but was cancelled because of a fault with a spacesuit.

The fruit of co-operation between NASA and the European Space Agency, the Skylab laboratory was a large cylinder a little more than 13ft in diameter and 16ft long. It was placed in the Shuttle's hold and linked to the latter's occupied area by a pressurised tunnel. In this 1982 picture, Vice-President Bush is framed on the right by Owen Garriott, the astronaut who spent 60 days on board Skylab in 1973, and on the left by the German Ulf Dietrich Merbold, who was with Garriott on Spacelab 1 in 1983.

On 28 November, 1983's missions ended with STS-9 (on *Columbia*), which witnessed the first space flight by a German astronaut (Professor Ulf Dietrich Merbold), as well as the entry into service of the Spacelab. This was an independent laboratory, a kind of mini-station stowed in the Shuttle's hold. Without a proper orbiting station and lacking the funds to match the Soviet Salyut, NASA had hit on the idea of encouraging other nations to take part in a new space project. The United States had known right from the Shuttle's conception that it would be very expensive, while the reductions in NASA's budget meant that it would not be possible to get involved with any other space systems. By signing an agreement with the European Space Agency in 1973 NASA was able to make substantial savings, while giving other nations the chance to benefit from a convenient, modern launcher.

Spacelab was to be formed of one or two pressurised modules in which scientists could work, linked by a tunnel to the Shuttle's living quarters. Equipment could be carried in the shape of the space pallet, fitted with materials or instruments, as well as measuring devices (the IPS, or Instrument Pointing System). Constructed by the German firm of Erno-VFW Fokker, with the participation of the British Aerospace Corporation for the pallet system and of Dornier for the IPS, Spacelab was a modular system whose total

length did not exceed 23ft. The first Spacelab, LM1, was given to NASA in exchange for flights by European astronauts, but the second, LM2, was bought by the American agency. The two laboratories were in use until 1998, by which date it was judged that they were no longer serving any useful purpose, because of the entry into service of the International Space Station. Nonetheless, the pallet system is still in use.

The Shuttle recommenced its flights in 1984 with mission STS-41 B. With this mission, designations were changed, STS-11 becoming STS-41 B. The figure '4' was the reference number of the fiscal year and the '1' indicated the base used for the departure (1 for the Kennedy Space Center and 2 for the Vandenberg base, still unused to this day). Finally, the letter 'B' indicated the mission's order in the programme. *Challenger*'s fourth mission, STS-41 B began on 3 February 1984, but the two satellites it carried could not be put into the correct orbit – this was undertaken by mission STS-51 A. The public found the EVAs performed by astronauts McCandless and Stewart particularly memorable, as they were able to move around freely a few metres from the Shuttle using their MMU (Manned Manoeuvring Unit) equipment. This was an independent unit fitting around the astronaut's backpack and fitted with 24 small thrusters controlled from a kind of armrest. The MMU allowed one or more members of the Shuttle's crew to go out and repair a satellite and to bring it back to the hold when the crane could not be used. Ultimately this sophisticated piece of equipment was little used, as it was judged to be too risky in operation. But for a few minutes, McCandless was the Earth's first human satellite. On its return to Earth the orbiter made the first use of the new airstrip at the Kennedy Space Center instead of landing at Edwards.

From April to November 1984, four further launches were carried out (missions STS-41 C, STS-41 D, STS-41 G and, with the new fiscal year beginning in November, STS-51 A). In the summer a new Shuttle, *Discovery*, was put into service. This was considerably lighter than *Challenger* and *Columbia*, as it no longer carried the equipment required to take a Centaur rocket stage in its hold (*Columbia* and *Challenger* could carry such a rocket with the object of launching various craft into space, but as it had never been tested NASA decided to abandon Centaur on all further Shuttles). The ejector seats – whose use was deemed risky – were likewise removed. *Atlantis*, the last of the Shuttles, joined the orbiter fleet in September 1985.

All the ensuing missions were a success, with closely spaced launches throughout 1985. During the month of April alone, two missions were carried out by *Discovery* and *Challenger*, and the timetable was equally full in October with two further launches. From a technical point of view, the flights were a complete success. The Shuttle stayed only a few days in space – rarely more than eight to nine – but long enough to put satellites into orbit, maintain them and sometimes bring them back to Earth in the event of a failure. Some of the missions had a purely scientific purpose. On paper, the orbiter seemed to present the ideal solution. One hundred and twenty-five people were carried in the 'space truck' between 1982 and 1986: a record! These successes allowed NASA to feel a little better about the lead taken by the Soviets in the space-station field.

The STS-71 mission's family photograph. This mission not only made a rendezvous with the Mir station in June 1995, but also took Spacelab up in its hold, where this shot was taken. In the picture are five of the Shuttle's crew and the crews of Mir 18 and Mir 19, the latter having just taken over from the former. This is the LM2 Spacelab module, the second to be built.

Bruce McCandless was the first astronaut to test the MMU (Manned Manoeuvring Unit) that allowed astronauts to move around independently, without being directly attached to the Shuttle. Shaped like an armchair, it was a propulsion system that fitted around the dorsal equipment. Two armrests carried the control. On Earth this device weighed 325lb, but it was relatively simple to use. Propulsion was provided by two tanks filled with liquid nitrogen, which would last for six hours. While the MMU allowed an astronaut to 'walk' around, it was of very limited use for working in space. The gases ejected could cause damage, and after the 1986 *Challenger* disaster this accessory was among the items judged to be too dangerous and was withdrawn.

The Buran Shuttle: a storm in a teacup

Created in answer to the American space Shuttle, the Soviet Buran ('Snowstorm' or 'Blizzard') orbiter resembled it closely. The aerodynamic form was indeed identical, as was the general design. However, the Soviet Shuttle differed on several points: the Energia rocket that propelled just the orbiter (which did not have any main engines) was composed of four boosters running on liquid fuel (not two solid-fuel boosters). This extremely powerful rocket was not reusable and could put a payload of up to 100 tons into orbit. Unlike the American Shuttle, Buran was equipped with engines allowing it to fly in the Earth's atmosphere. It could also carry a payload five tons greater than the American orbiter – 30 tons as against 25. Finally, the Soviet vehicle was capable of making unmanned, automated flights, something that was quite impossible with NASA's Shuttle. There was room for four cosmonauts on the main deck and each had an ejector seat. On scientific missions it was possible to add up to six more (non-ejectable) seats on the lower deck. Heat-protective tiles made from synthetic quartz and organic fibres, covered in places with carbon, ensured the structure's protection on re-entry.

The Buran programme began in 1971, but it was not actually until 1984 that the first OK-GLI Shuttle was constructed. It was intended to validate the concept and was used only for experimental flights in the atmosphere. For this purpose it was fitted with supplementary jet engines. Five other orbiters were constructed for various tests – thermal, acoustic, electronic – before, in 1986, the first operational Shuttle emerged from the Korolev plant in Moscow. It was taken to Baikonur on the back of an Antonov AN-225 heavy transport plane (the largest plane of its type in the world, and unique). A three-mile hard runway had been built at the launch pad for the Shuttle's return. The OK-1.01 vehicle was attached to an Energia rocket and moved into the vertical position.

On 15 November 1988, Buran took off, unmanned, and made two orbits of the Earth. It then made a fully automated re-entry and landing. This was an outstanding technical achievement, as a fully automated flight was very risky – NASA had flatly given up on this idea for its own orbiter. But Buran would make only this one space flight. Severely hit by the economic crisis that followed the collapse of the Soviet Union, the programme began to take on water. The second Shuttle (OK-0.02, alias Buria or Ptichka) was 95 per cent complete and was due to take off in 1993 to go up to Mir. In the end, however, it was to be the Space Shuttle that carried out these missions.

Buran and Energia were abandoned in 1993. At first the two Shuttles were stored in atmosphere-controlled conditions at Baikonur's MIK complex, but, surplus to requirements, and the property of Kazakhstan, they were eventually moved to other buildings. In 2002 the hangar where Buran was stored suffered storm damage and the Shuttle was crushed when the roof collapsed. Only the second Shuttle survives.

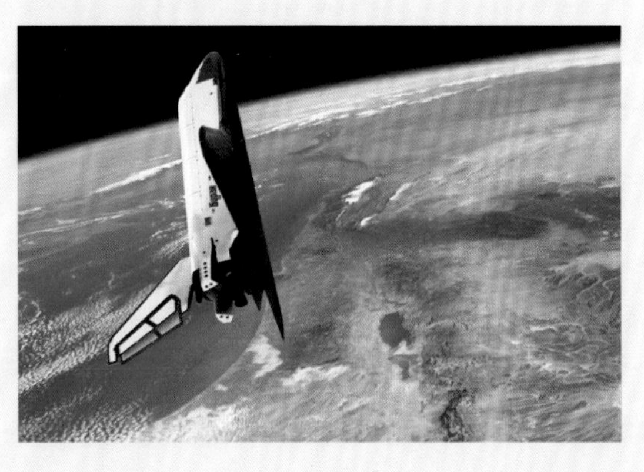

A fine picture of Mir in September 1996. The station is in its definitive form, with its seven modules. Note the size of the solar panels, which take up a considerable amount of space. They supplied all the on-board electricity and were of vital importance for carrying out all the scientific experiments.

Mir, a decisive step

After a rather lean period, the Soviets decided to make a big impression with Salyut's successor. In Russian, Mir means both 'peace' and 'world'. This DOS-17KS-type station was another leap forward in space exploration. Unlike Salyut, it was assembled in space in several stages, like a giant Meccano set, with the addition of several modules between 1986 and 1996. The same solution was subsequently adopted for the International Space Station. Mir made its 'paper' debut in 1976, when it was decreed that the project should be adopted. With Chelomei's Almaz military stations having finally been set aside, the concept of Mir was able to take shape quite rapidly. Entrusted to the NPO Energia group, its construction was in fact contracted out to KB Salyut (formed in 1976, this sub-contractor to NPO Energia became independent in 1989). It had docking points in the standard locations at the front and rear of the station – one for a Soyuz craft and the other for a Progress-type supply ship – but Mir's originality lay in its central docking module that allowed it to link up with as many as four other systems, either spacecraft or other modules derived from TKS craft but fitted out as laboratories and weighing close to 20 tons.

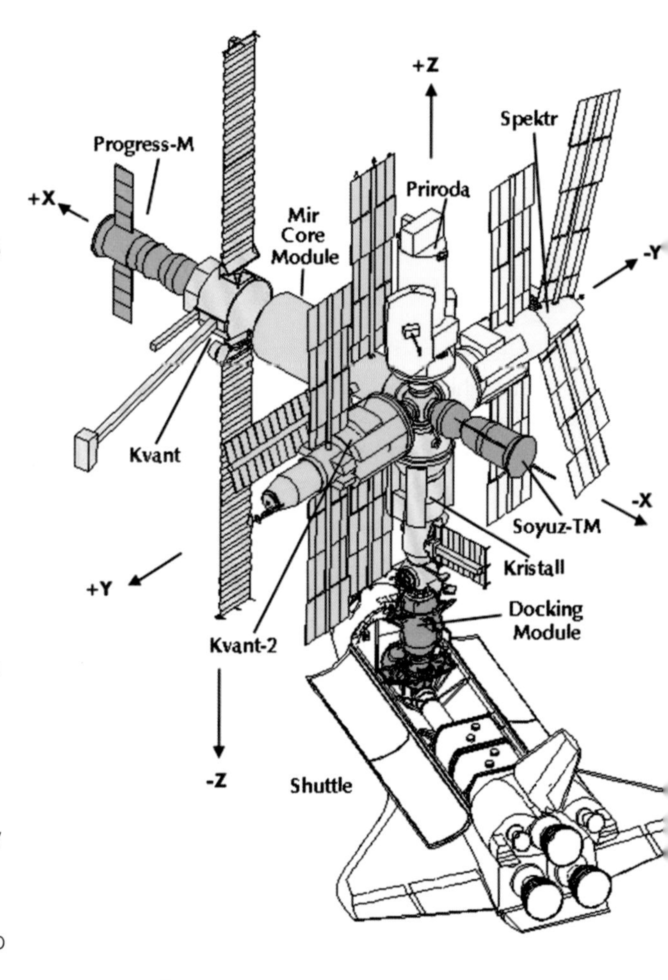

The main module (or base block) of the Mir station came from the Korolev plant. It was an improved version of Salyut with a rear docking port and a sphere (the transfer compartment) providing docking points for five other modules or craft of the Soyuz type. This first element was Mir's backbone, all other modules simply being extensions.

Work was well advanced when the Buran Space Shuttle programme was given priority in 1984. For budgetary reasons, Mir almost did not get completed. It needed all the authority of the Minister of Defence and Space, Valentin Glushko, to put Mir back on centre stage. Tests had revealed, however, that the main module was a ton heavier than planned. Even after all the scientific equipment was removed (it would be taken up to the station separately on a Progress craft), it remained too heavy for a Proton rocket to insert itself into a 65° orbital angle. Eventually the engineers decided to send the first module on a 51.6° orbit – like Salyut – even though this would not put the station high enough to photograph the Soviet Union in its entirety. The work was delayed and it became clear that the planned digital computer system would not be ready in time. The station therefore had to be fitted initially with the previous generation Argon analogue system. On 20 February 1986 a Proton rocket lifted off from Baikonur with the first central element of the station, which was put into an orbit with a perigee of 240 miles and an apogee of 245 miles. This module was to serve as the cosmonauts' living quarters, and also as the control and communication module. Unlike Salyut this module was poorly provided with scientific equipment, as this was to be sent up later on other modules.

With its six docking points, the central module was the backbone of the whole programme. It was divided into several distinct parts. The work compartment also enclosed the living, ablutions and rest area. It was composed of two connected cylinders with a maximum length of 25.16ft and a maximum diameter of 13.78ft. To make it easier for crews to orient themselves, the engineers had had the floor, walls and ceiling painted in three different colours. The spherically shaped transfer compartment with its six docking points – with four of them arranged at 90° to each other – allowed other modules to dock and/or to receive other craft. The intermediate compartment was originally used as a tunnel linking the work compartment to the rear docking-point airlock (it later joined the Kvant module to the work compartment). Finally, the last 'assembly' compartment contained the fuel tanks for the main engines used for changes of orbit and for the Vernier engines used for yaw, pitch and roll movements.

After various tests, the station became operational on 6 March 1986. It actually passed within six miles of the still-functioning Salyut 7. On 13 March its first crew, consisting of Leonid Kizim and Vladimir Soloviov, arrived on Soyuz T-15. For the first time in the Soviet space programme, they were presented to the international press the day before departure, an action which demonstrated that the Soviets were now keen to show a degree of openness and had less reason to hide their civilian space activities from the eyes of the world. Soyuz T-15

Mir's first permanent crew consisted of Leonid Kizim and Vladimir Soloviov. They took off on board Soyuz T-15 on 13 March 1986, barely a month after the launch of Mir's first section. An unusual aspect of this mission was that the crew also visited Salyut 7, 2,500 miles from Mir and in a lower orbit. To save Soyuz's fuel, Mir's orbit was lowered by more than half a mile so that a gentle foot on the accelerator brought it closer to Salyut 7. When Salyut was within 1,500 miles, Soyuz T-15 undocked from Mir and made the 29-hour journey to Salyut. Everything went perfectly right up to the return to Mir on 26 June.

docked at Mir's front end, so as to leave the rear port free for the Progress supply ship. However, the operation was not straightforward, as only Mir's rear port was fitted with the same Igla guidance system as the Soyuz vessel (the front end had the new Kurs system fitted): they therefore had to approach from the rear, then once within sight, turn the craft around manually and dock at the front. Once they were on board, Progress 25 joined them on 21 March. Various resonance tests were undertaken to test the structural strength of Mir and the two docked vessels; Progress 25 then left the station and was replaced on 26 April by Progress 26. Kizim and Soloviov then took Soyuz T-15 and joined Salyut 7 on 6 May. Salyut 7 was in a very poor state, but the two men repaired what they could and carried out two EVAs.

The Russian mission continued with its exceptionally varied activities. Soyuz T-15 departed from Salyut 7 with Kizim and Soloviov on 25 June and rejoined Mir the next day, with 880lb of recovered scientific equipment. In the meantime Mir was visited by an unpiloted Soyuz TM-1 designed to test the Kurs guidance system, and Progress 26 uncoupled and was burnt up as planned on re-entry. For their part, Kizim and Soloviov returned safe and sound to Earth on 16 July. The mission had been a complete success. It was the first time that two cosmonauts had travelled from one space station to another, undertaken two spacewalks and carried out nearly 150 experiments. In just a few weeks, the Soviets had managed a great technical feat. The Americans had no reply, as they had just suffered a tragedy that was to shake the very foundations of the Space Shuttle concept.

The Mir station in its initial configuration with the Core element. No other module has yet been attached – these would arrive over time. Three of them would be operational before the 1990s (Kvant-1, Kvant-2 and Kristall), but it would be another five years before Spektr's arrival. The reasons for this hold-up were simple: with the fall of the Berlin Wall communism was in a state of collapse. Disorganisation reigned as the former Soviet Union fell apart (it became the Russian Confederation in 1991), and the Mir programme was seriously delayed.

The success of Proton

Proton was designed by Chelomei, Korolev's direct competitor. Chelomei wanted to develop a universal rocket (UR, for *Universalnii Raketi*) capable of being used both as a launcher and as an intercontinental missile. He had already built the UR-200 rocket, but this, with its modest power, could not meet the ambitious needs of the Soviet space programme. Project UR-500 thus took shape, but Chelomei ran into the problem of providing engines. There were some made by Glushko of OKB-456, but these were risky because of the hypergolic mixture used as fuel, and they had been turned down by Korolev for his N1; but, in the absence of anything better…

The vehicle designed was a fairly big (688-ton) two-stage missile, capable of carrying a 20-ton military payload. But Krushchev soon got cold feet and the UR-500 was never operational as a weapon. Nevertheless, in 1965 a UR-500 put a Proton satellite into orbit – and the name stuck. It was the rocket's first firing and its success was encouraging. What was less encouraging was the fact that it was one and a half times bigger than Korolev's R-7, but carried a payload only 25 per cent greater. A further stage therefore had to be added.

The civilian rocket was designated 8K82 in the Soviet system. The first stage (Proton K-1) had six RD-253 engines encircling the main body, each with its own fuel tank containing UDMH (unsymmetrical dimethylhydrazine); the oxidiser (nitrogen tetroxide) was contained in the main body of the rocket. These engines were not detachable boosters as they might appear. The liquid mixture they burned is a toxic pollutant, especially if it gets into the water supply. Furthermore, it is still being used, provoking numerous 'green' protests in Russia. The 8K82 rocket's second stage

(Proton 2) was an enlarged version of that on the UR-200. Capable of putting into orbit a payload of up to 8.3 tons, this launcher was followed by a bigger version, the UR-500K or 8K82K. Having heavier and more-powerful first and second stages and topped by a third stage, this was able to carry a payload of 20 tons.

After Krushchev's departure – he was replaced by Brezhnev – Korolev was once again in favour and the head of OKB-1 would turn the situation to his advantage. The UR-500 was certainly Chelomei's work, but the lunar programme would be headed by Korolev. He now put forward his circumlunar probe, Zond, which he intended to put on a Proton rocket. But to send the probe to the moon, Korolev needed to add a fourth stage. So he brought his touch to the rocket by adding the final stage from the unsuccessful N1, which he had designed. Thus equipped, the Proton rocket began to assume the definitive form in which it would fly. However, the development was difficult and protracted, with a fairly high failure rate until the early 1970s. It would be improved over the course of the ensuing years until it became one of the most reliable and economical rockets of them all. Indeed, the Proton was the preferred method of putting into orbit all the elements of the Salyut and Mir stations, as well of the ISS. Today, the Proton M version (8K82M) is the most advanced of the family. Lighter, more powerful and less polluting, it is the most frequently used commercial launcher.

Technical details

Name: *Proton 8K82K*

Type: orbital launcher

Country: Soviet Union

Manufacturer: Chelomei

First launch: 16 November 1968

Last launch: 12 July 2000

Power on lift-off: 8,847kN

Engines:

– First stage:
6 RD-253-11D48 rocket engines giving 1,050 tons of thrust for 124 seconds

– Second stage:
4 RD-210 rocket engines giving 240 tons of thrust for 206 seconds

– Third stage:
1 RD-0212 rocket engine giving 64 tons of thrust for 238 seconds

Total weight on lift-off: 682 tons

Height: 164ft

Empty weight: —

Payload: 43,560lb

The Proton rocket was used to put all of Mir's parts into orbit (except the docking module for the Shuttle). This photograph – probably taken at the end of the 1960s – shows the launcher as it was originally designed, to take the Zond (Soyuz) 7K-L1 craft to the moon.

The crew of mission STS-51 comprised, from left to right and top to bottom, Ellison S. Onizuka, Sharon McAuliffe, Gregory Jarvis, Judith Resnik, Michael J. Smith, Francis 'Dick' Scobee and Ronald McNair.

The *Challenger* explosion

It was exceptionally cold on 28 January 1986 at the Kennedy Space Center's launch pad 39A. *Challenger* was preparing for its tenth mission. On board the orbiter were seven crewmembers, one of whom was a primary-school teacher. Sharon McAuliffe had been selected from among more than 11,000 candidates to become the first teacher in space. NASA had put together a teaching programme (the TISP, Teacher In Space Programme) to revive enthusiasm for space exploration in future generations. McAuliffe was also the first truly civilian participant in the space programme, and naturally this had mobilised the media. *Challenger* was commanded by Francis Dick Scobee and piloted by Michael J. Smith. With McAuliffe was another woman, Judith Resnik, a scientist, as well as three other specialists: Ellison Onizuka, Ronald McNair and Gregory Jarvis. The Shuttle was to put a TDRS satellite into orbit, follow the tail of Halley's Comet

The launch pad at the Kennedy Space Center was partially covered in ice. During the night before the launch, de-icing operations had been carried out to remove the ice that had accumulated on the Shuttle and the tower. Rockwell International, the Shuttle's builder, had warned NASA that ice might cause damage to the Shuttle's skin during its rapid ascent through the atmosphere, but while several inspections had certainly taken place, the countdown was not halted by the weather conditions.

At *Challenger*'s launch, after the successive ignition of the main SSME engines and the boosters, dark grey smoke was noticed emanating from the lower part of the right booster, where it was attached to the external tank. The smoke, nonetheless, dissipated very quickly. As *Challenger* began to climb, nothing out of the ordinary occurred. All was proceeding normally when, less than a minute into the flight, the cameras tracking the Shuttle clearly showed an incandescent plume coming from the booster, away from the nozzle. The flame was playing against the side of the external tank.

None of the on-board computers could detect a leak in the boosters. This solid-fuel rocket is a bit like a match: once lit, it doesn't stop until it is fully burnt up. But on the external tank the sensors indicated an abnormal fall in the pressure of the liquid hydrogen in its lower half. In contact with the air and the flame, the liquid hydrogen poured out and caught fire, followed by the liquid oxygen. The tank burst open and the boosters separated and continued in uncontrolled flight, while the Shuttle came away, its own tanks exploding in turn. The whole of the front section broke off and fell towards the ocean.

Jay Greene's face expresses more than incredulity. Greene was NASA flight director from 1982 to 1986. NASA imposed a complete news blackout on the accident in the ensuing hours and no one in the control room gave interviews. Greene's role was limited to following the launch and flight procedures. Nothing out of the ordinary had been detected during this phase, as the defective booster was not fitted with sensors. Greene, who had been one of Apollo flight director Gene Kranz's right-hand men, resigned after the accident.

using 'Spartan' instruments, and carry out various experiments.

The mission had been postponed several times, largely because of weather conditions, including those prevailing at emergency landing strips abroad. At 11:38, New York time, the main engines began to roar. At this point, with the boosters not yet in action, it was still possible to abort the lift-off; but 6.6 seconds later the huge solid-fuel rocket engines thundered into life. Unlike an engine running off a liquid-fuel mixture that can be shut down by cutting off the flow of fuel, a solid-fuel engine cannot be stopped. As with a match bursting into flame, you have to wait until all the powder has been consumed before it will stop burning. Just before lift-off, a puff of brown smoke was noticed on the right-hand booster, at the point where it was attached to the external tank. But it was too late in any case to halt the launch, as the explosive bolts holding the Shuttle to the pad had been blown, freeing it for lift-off. The smoke had been coming from a slow but definite combustion escaping from the joints separating the two sections of the booster. The Shuttle took off as normal, turned, and rose into the blue sky. At T plus 37 seconds, strong wind shear was experienced, obliging the on-board guidance and control computer to stabilise the Shuttle. It was not a serious problem, but the forces endured by the orbiter's structure were the highest that had been recorded up to then. At the same time, the power of the SSME and booster engines was reduced to avoid excessive speed in the dense layers of the atmosphere. After a few seconds the engines returned to full power to provide the maximum thrust.

At T plus 60 seconds, a long thin flame became clearly visible on the right booster. The pressure in the defective solid-fuel rocket fell rapidly, proving at the same time that there had definitely been a leak. The thin flame was getting longer. The aerodynamic flow directed the flame towards the external tank, burning up the booster's fixings. It pierced the tank, releasing hydrogen that immediately burst into flames beneath the Shuttle. Everything then happened very quickly, in one or two seconds. The right booster, missing one of its fixings, now rotated, broke its upper fixings and ruptured the lower part of the external tank where the liquid oxygen was concentrated. Transformed into a fireball, the tank collapsed, enveloping *Challenger* in an incandescent mass. At almost mach 2, the Shuttle was subjected to exceptional stresses and began to break up 73 seconds after launch; the two boosters continued their uncontrolled ascent for a further 37 seconds. Contrary to what might be expected, there was no explosion; no detonation was noticed. The powerful forces and aerodynamic strains broke the Shuttle up into several pieces. The cabin, as well as the (more robust) SRB solid-fuel rockets fell into the ocean and were destroyed on impact with the sea at over 200mph. The tragedy lasted until another ten seconds or so before the impact, as it appears that some members of the crew were still conscious as the pieces fell towards the sea.

For many a long minute, the TV cameras remained pointing up at the sky, as if it was just a passing incident and the Shuttle might continue its climb.

Can the crew be saved?

There has been much debate over the Shuttle's lack of any proper means of escape. In fact the first Shuttles were fitted with ejector seats identical to those used in the Lockheed SR-71, but only the pilot and the commander could eject, the rest of the crew being confined to the main deck or behind, without any means of evacuating the craft. The ejector seats had really been designed for the early tests (with two crew members), to be used during the gliding descent at subsonic speeds, and then only if the landing attempt failed. Ejecting from the Shuttle during the tricky lift-off stage would have been very dangerous, as the pressure during the ascent is such that a human being would be unlikely to survive. The only feasible solution would have been to provide the Shuttle with an ejectable pod (like that on the Rockwell B-1B Lancer). Studies were carried out, but the excessive cost of such a system, not to mention the extra 5.4 tons, did not seem worthwhile in relation to the probability of an accident.

NASA was confident of its orbiter (which currently has a 99 per cent success rate) and adopted the same view as the majority of the builders of civil airliners, which are also lacking any means of in-flight evacuation. However, it has to be admitted that travelling on an Apollo-type craft is less risky than a Shuttle flight.

After the *Challenger* accident a parachute escape system was developed, but it could ensure the crew's survival only under certain conditions and would have been of no help in that particular accident, still less so in *Columbia*'s case. In the end, when considering safety on board a rocket-launched spacecraft, it has to be understood that on lift-off, the crew and the engineers on the ground have no means of controlling it. Computers may monitor and manage the optimum trajectory, but if a rocket or a Shuttle should deviate too seriously during the first few minutes of flight the only solution is to destroy the craft.

A meticulous enquiry...

A presidential commission led by the former Secretary of State, William P. Rogers, was charged with determining what had actually happened. It naturally turned its attention to NASA and the booster manufacturer Morton Thiokol. The latter was aware of the Achilles heel of the notorious O-rings and had made NASA aware of this well before the disaster – flight STS-51 B of 24 January 1985 had had identical problems without actually leading to the joints' complete failure. One might draw some conclusions from the fact that the following day President Reagan was due to make his State of the Union address (the most important presidential speech in the US) and that *Challenger*'s launch would have made a good introduction. It could also be considered that the media attention surrounding Sharon McAuliffe was such that the Shuttle absolutely had to go (it had already been postponed several times). Finally, it should not be forgotten that flight STS-51 L was supposed to observe Halley's Comet, which was unlikely to wait for it! Whatever else may be said, the fact remains that the joints were faulty.

Technically, the cold weather at the launch site that day played a major role. It was -2°C (as low as -8°C around the Shuttle, thanks to wind chill), and ice was forming. Teams had been at work during the night melting it, but it had frozen again. The booster's O-ring joints were not intended to withstand such temperatures (there was a risk they could crack), which was well known to Morton Thiokol's engineers, who had warned their superiors about it. A discussion took place on the evening of 27 January between NASA's managers and representatives from Morton Thiokol. Finally, they decided to go for a launch despite the warnings. It was the same story from the Shuttle's builders, Rockwell, who were concerned about the ice, which might cause damage to its structure or the engines. But the manufacturer's recommendations were not interpreted as meaning that the lift-off should be cancelled. There was undoubtedly negligence, under media and indeed political pressure, and it cost NASA dear.

After the *Challenger* and *Columbia* disasters, Shuttle crews were to wear an orange pressurised suit: the LES suit (introduced after 1986), then the more modern ACES suit, which is still in service. At the same time, a rapid Shuttle escape system (the CES, or Crew Escape System) was developed, but was usable only when the Shuttle was in gliding mode during its return.

The members of the Rogers Commission arrive at the Kennedy Space Center. The commission produced a very complete 225-page report, but the physicist Richard Feyman created ripples when he showed that the technical failures had also been human failures. His comments were eventually added to the report.

The orbiter faces an uncertain future

After *Challenger*, the Shuttle's future appeared somewhat compromised. Beyond the accident and the responsibilities flowing from it, there was a clear realisation that the Shuttle was much more expensive than had been foreseen and that the cost per kilo launched into space had in no way been reduced, despite a busy programme. The STS had been intended to make flights more or less weekly at a cost no higher than one-tenth of a normal 1970s launch. In fact, the maintenance of each Shuttle had blown these figures out of the window and, in particular, considerably reduced the number of launches. The US Air Force wanted to withdraw from the programme, and at NASA some people were wondering whether it wasn't time to retire the orbiters. Grounded for two and a half years, the Shuttle seemed to have little further reason for its existence. Simultaneously, the Freedom orbital station, initiated by Reagan in 1984 and due to enter service in 1994, was in a bad way, as were the Hubble and Galileo programmes.

Back at the head of NASA, James Fletcher tried to breathe some new life into things. The task was arduous. The administration had become very bureaucratic and managers and engineers needed their confidence rebuilding. The worst thing was that no programmes had been established for the future. NASA seemed to be drifting without purpose and to be incapable of developing any big ideas. Early on, in July 1987, Fletcher set in motion the building of a replacement Shuttle for *Challenger*. The *Endeavour* orbiter was not built like the other Shuttles: it was put together from parts originally intended for repairing the others. Nevertheless, *Endeavour* received much new equipment, which was later retrofitted to the other vehicles. For example, a parachute-brake was fitted, reducing the braking distance from 2,000ft to 1,000, and the avionics were considerably improved. At the same time the vessel could now carry out very much longer missions (up to 28 days). This Shuttle did not go into service until 1992, but in the meantime it was *Discovery* that recommenced flights, on 29 September 1988. Mission STS-26 took place without any hitches. After *Challenger*, NASA had decided to go back to the old mission numbering system. On board the Shuttle, the five astronauts now wore pressurised suits. The TDRS satellite was successfully put into orbit and the crew returned to Earth four days later. There was great relief all round. Flights could now go ahead as normal.

The concept of the Freedom station came from NASA's sketchbooks just as the space Shuttle was becoming operational. It seemed obvious that the orbiter was perfect for use as a space bus or truck up to a permanently manned space station. James M. Beggs, NASA's administrator, said that this was the 'next logical step'. It was President Reagan, in 1984, who gave the green light. Studies focused on quite a different configuration from that used in the future ISS, with modules located at the centre and solar panels projecting from them. Eventually, because of the cost and Congress's unwillingness to release the necessary funds, Freedom lost its interest. The new Clinton administration preferred to consider a co-operative effort with the Russians.

Ex-astronaut Richard H. Truly, NASA's eighth administrator, gives an official speech at the presentation of the new Space Shuttle in May 1991. Truly was the first trained astronaut to head the administration (from 1989 to 1992). The construction of the Shuttle *Endeavour* began in 1987 as a replacement for *Challenger*.

On 29 September 1988, STS-26 was the 'return to flight' mission, the first since the loss of *Challenger*. On board *Discovery*, the whole crew now wore a pressurised suit during take-off and entry into orbit. STS-26 was not just a simple test mission; they also had to put a satellite into orbit before returning to Earth four days later.

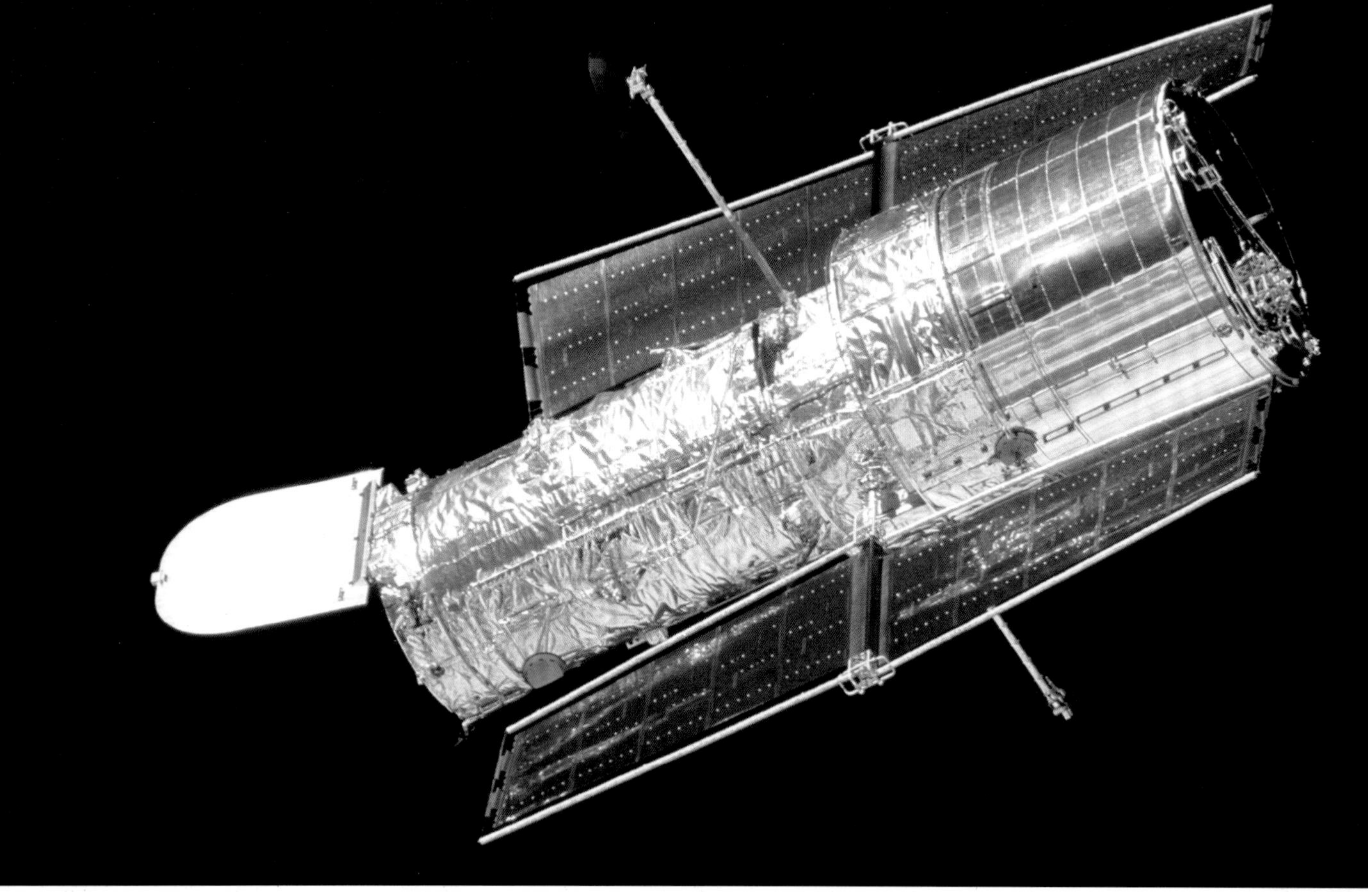

The Hubble telescope as it appeared to the members of mission STS-82 in 1997, on the second maintenance trip. The flap protecting the front lens can clearly be seen. The main mirror (opposite page, during assembly of the telescope) was designed by the firm of Perkin-Elmer and required hours of polishing to achieve a perfect surface. Kodak took on the job of providing a second, replacement mirror, and Lockheed built the structure of the telescope. Perkin-Elmer made use of an entirely automated polishing process, but a mistake in the calculation of the mirror's curve required the installation of corrective mirrors in 1993.

The case of Hubble

The Hubble space telescope weighs 11 tons, is 43.3ft long and orbits at an altitude of around 360 miles. It was put into orbit by *Discovery* on 25 April 1990, during mission STS-31. The decision to send a telescope outside the Earth's atmosphere was an old idea, born in the days after World War 2. It was the astronomer Lyman Spitzer who laid the foundations for the observation of distant planets and stars from space. It is a well-known fact that the higher one goes, the easier such observation becomes, thanks to the absence of pollution and the atmospheric layers. In space, resolution is even better and it is also possible to observe infrared and ultraviolet rays – barely visible on Earth, as they are partially absorbed by the atmosphere. The space telescope programme was soon integrated with the space programme, and space astronomy took off well before Hubble, with the sending into space of various instruments from the beginning of the 1960s. But in order to make the concept worthwhile it was necessary to be able to carry out regular maintenance, and, in the early 1980s, the Shuttle arrived at just the right moment. However, any future space telescope would require funding, and once again the problem of costs almost scuppered the project. The American Congress made cuts in the initial budget then went as far as turning it down. Pressure from the scientific community caused them to revise their decision, particularly as the European Space Agency had agreed to contribute to the costs of the telescope, whose expensive main mirror was nevertheless reduced in diameter from 3.0m to 2.4m.

At the start of the 1980s the space telescope was christened 'Hubble' in honour of Edwin Hubble, the astronomer who had discovered that the universe was expanding. The construction was delayed and the cost of the programme increased, but by the beginning of 1986 everything was ready. Unfortunately, the *Challenger* disaster grounded the Shuttles and Hubble had to wait until 1990 to get into orbit. Right from the start, the pictures it sent back seemed rather blurred. Indeed, the pictures were no better than those taken by a terrestrial telescope. This was more than disappointing, as Hubble had eventually swallowed up five times its original budget – nearly two billion 1999 dollars! Hubble became the focus of media attention... and the target of much scoffing and mockery. Quite soon it was realised that the mirror did not have the correct curvature, thanks to an error in the original calibration. In effect Hubble could see, but was short-sighted. Since no adjustment could be made from the ground, it was decided to provide the telescope with correction mirrors. It had in any case been planned to carry out regular maintenance, even if this was only to raise the telescope's orbit, as without an engine it was

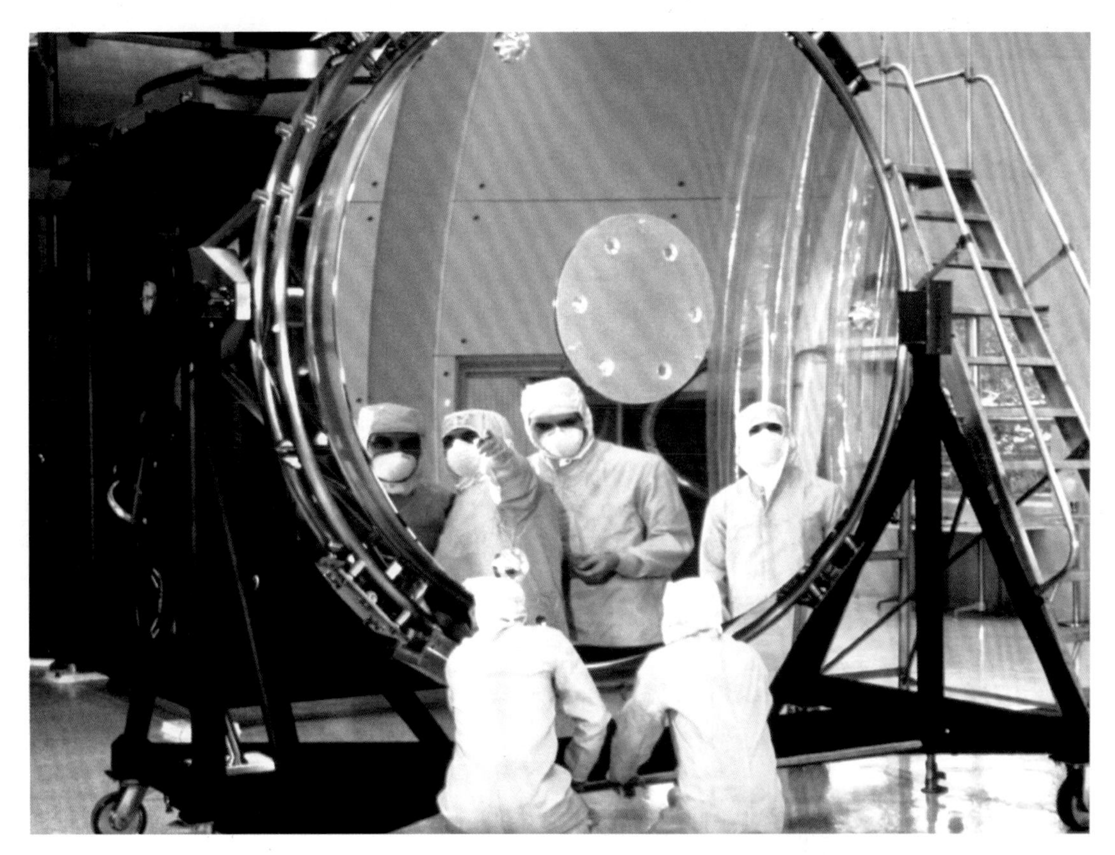

Mission STS-31, carried out by *Discovery* on 24 April 1990, took the Hubble telescope into orbit. It required four other missions to maintain the telescope in working order. Today, Hubble's future still remains uncertain, such is its demand for maintenance.

slowly but surely descending back towards the Earth, with the consequent risk of being burnt up on re-entry.

It was 1993 before mission STS-61 was able to resolve the problem. The Shuttle *Endeavour* and its crew (including the Swiss Claude Nicollier, making his second flight) were about to achieve a remarkable technological exploit. After recovering the telescope using the robot arm operated by Nicollier, astronauts Musgrave and Hoffman carried out an initial EVA. There would be a further four during the 11 days the mission lasted. They would crush all previous records, with between six and seven hours for each EVA. These were all undertaken to install Hubble's 'spectacles', the COSTAR system (Corrective Optics Space Telescope Axial Replacement), as well as putting in a new camera and changing the electronic modules. Two thousand tools were used to do all this. Though it was a complete success, Hubble nevertheless required further maintenance in 1997 (with STS-87), 1999 (with STS-103) and 2002 (with STS-109). NASA's orbiters were by no means unemployed. From 1998 to 2000, just before the flights to the ISS, the Shuttles carried out 71 missions. On 29 October 1998, STS-95 took astronaut John Glenn on his second space flight. The first American in space was then 77 years old and is, to this day, the oldest person to have gone into space.

JPL
FRAME

A magnificent picture of Saturn, taken by Hubble in 2005. You can clearly see the planet's polar auroras. This photograph has been manipulated, as the auroras are visible only under ultra-violet light. Only Hubble has the equipment to take this kind of shot.

This very fine picture was taken during mission STS-61, when Hubble's 'myopia' was corrected by the installation of Costar, a correction package of five mirrors. The telescope has been brought to *Endeavour*'s hold and some of the crew are working on it. Note how the arm is being used to carry the astronauts to where they are working. This was done after the scrapping of the MMU system.

Frenchman Jean-Loup Chrétien was part of the Aragatz mission and took off on board Soyuz TM-7 on 26 November 1988. This was his second trip to a station and on this occasion he carried out an EVA of almost six hours.

In pursuit of Mir

The number of missions, however, masked a reality. In the field of space stations it was becoming impossible for the Americans to keep up with the Soviets, who had already begun to send a first module up to Mir. This was Kvant-1, an astronomical study module. It was intended to dock with Salyut 7, but with its development delayed it was not possible to get it ready in time. This turned out to be a piece of luck, as Kvant-1 was in any case too heavy to be carried by a Proton rocket into the correct orbit planned for Mir. As the first section of Mir could not be sent into a 65° orbit either, Kvant-1 found its *raison d'être* once the engineers had agreed on the lower 51.6° angle of inclination. Lacking engines, Kvant-1 was towed by a second TKS-type craft, specially adapted for this mission. It had to bring the module to Mir, separate from it and then self-destruct on re-entry into the atmosphere. Besides scientific equipment, the module also carried gyroscopes. But Kvant-1 should have docked at the rear end of Mir, exactly where the Progress craft docked. With its 1,400cu ft bulk, Kvant-1 could certainly fit between Mir and Progress, but it would need to have pipes added to pump fuel from the supply ship to Mir's tanks. Again, this would add to the weight. Eventually, it was decided to use a modified Proton rocket to transport the heaviest load ever carried by the launcher (22.4 tons for Kvant-1 and its tow vessel).

Kvant-1 lifted off on 31 March 1987 but was unable to dock properly. Cosmonauts Yuri Romanenko and Alexander Laveikin – who had arrived on 7 February aboard the first TM ('Transport Modified') series Soyuz – carried out an EVA to ascertain the state of the docking systems. On Kvant-1's airlock they found a piece of a plastic bag (or material according to some sources), almost certainly left there by a technician, which was preventing the hook-up with Mir. It was during this EVA that an incident occurred. As he was moving towards the module, Laveikin noticed an abnormal reduction in his spacesuit's pressure. Romanenko and the engineers on the ground heard him shout: 'My pressure's falling, the pressure's falling!' As he turned round, Romanenko saw that Laveikin's pressure regulator was in the minimum position. He quickly reached out and reset it to normal – on exiting Mir the regulator lever had got caught in one of the airlock's handles. The incident could have ended with a serious accident. After such a fright, Laveikin had had some minor cardiac troubles, which can be dangerous in space, so he was replaced a few days before the end of the mission by Alexander Alexandrov, who arrived on board Soyuz TM-3 on 22 July 1987 with two other cosmonauts, Alexander Viktorenko and a Syrian, Mohammed Faris.

This vessel remained attached to Mir and a few days later it was Soyuz TM-2 that took Laveikin, Faris and Viktorenko back, with Romanenko remaining on board alone. He would end up with a flight time of 326 days. In the meantime, he and Laveikin had undertaken two more EVAs (one of which was to install a third solar panel on Mir). When he returned to Earth on 29 December, Yuri Romanenko had grown by four inches, lost almost 20 per cent of his muscle mass and his bones had become fragile. Psychologically, he had coped well with the ordeal.

Thanks to the Soyuz TM craft, Mir was occupied on an almost permanent basis by the Soviets, as well as by guest astronauts. For example, Soyuz TM-5 took the Bulgarian

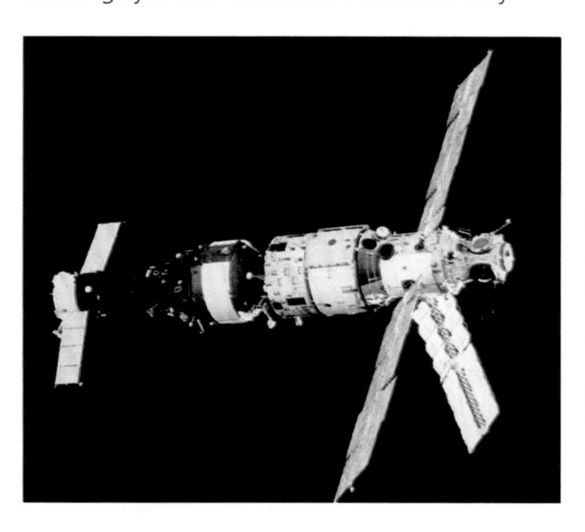

Mir is seen here docked with the Kvant-1, which is positioned between the crew's Soyuz vessel and Mir itself. Kvant ('Quantum') was an astrophysics module fitted with gyrostabilisers allowing the whole station to adjust its position without having to use its engines.

Laveikin and Romanenko were there when Kvant-1 docked. As the Soyuz craft had to move away to give up its place to Kvant, which would dock automatically, the crew had taken refuge in Soyuz TM-2 for safety. On the first approach, the Igla guidance system suddenly stopped working and Kvant overshot the station by 1,300ft. The second approach was carried out in manual mode, controlled by the crew. This time the docking worked properly, but the seal was not airtight and the electrical connections did not work. Some unidentified object was interfering with the connection. This turned out to be a plastic bag, which was discovered during an EVA. Unfortunately, after this EVA Laveikin began to suffer cardiac problems, which compelled him to return to Earth sooner than intended.

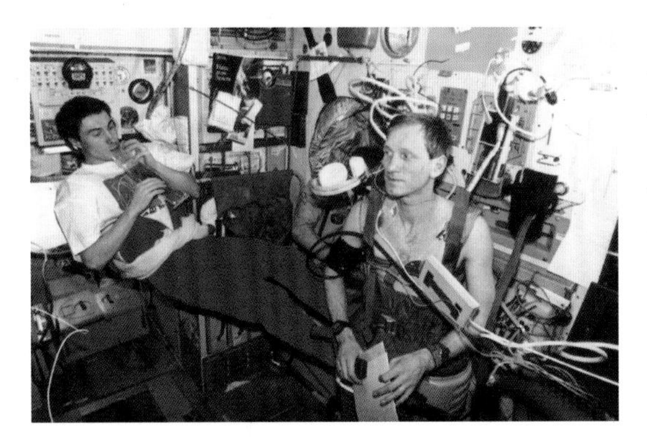

Alexander Alexandrov (no relation to the Russian cosmonaut of the same name); Soyuz TM-6 gave the Afghan Mohamad his first flight; and Soyuz TM-7 took off with the Frenchman Jean-Loup Chrétien on board, making his second flight and staying almost a month in orbit. The days of man in space were far from being over, with Mir growing ever larger.

The Meccano set in space

Following Kvant-1, the second module, Kvant-2, was launched on 26 November 1989. This huge element once again included scientific equipment, but also had living quarters for the crew's comfort (including a shower, water and an oxygen regeneration system). In its 2,100cu ft volume, Kvant-2 was fitted with thrusters for manoeuvring, six Gyrodyne gyroscopes, a metre-wide airlock for EVAs using a backpack propulsion system (like the American MMU), and two solar panels to increase Mir's electrical capacity – as well as the station's weight. The third module, Kristall, was attached on 1 June 1990, 11 days after its launch. Almost the same shape as Kvant-2, but slightly smaller, Kristall was a scientific module intended for astronomy, studying materials and living beings under weightless conditions and observing the Earth. It was connected perpendicularly at the front end of Mir, but its siting was changed several times using the station's crane, so as to accommodate the fourth module, Spektr. Mir was assembled and disassembled according to each mission's requirements, like a Meccano set.

Mir's story coincided with one of the big events of history in 1991, when cosmonauts Sergei Konstantinovich and Alexander Alexandrovich Volkov changed nationality while in space from Soviet citizens to citizens of the Russian Confederation! The end of the Soviet empire was initially a severe blow for NPO Energia, the industrial group responsible for building spacecraft and rockets, as the funds it was allotted were rapidly cut. Economically and politically, the new Russia was unstable and no one could predict Mir's future. For NASA, however, it was something of a relief. A new world emerged from the fall of the Soviet regime and the convulsions that followed. Glasnost finally allowed the Russians to unveil a whole range of their technology.

Time for co-operation

In 1993 profitable contacts were established between NASA and RKA, the new Russian space agency created in 1992. On the initiative of Vice President Al Gore and Russian Prime Minister Viktor Chernomyrdin, a huge international space station programme was launched that would become the ISS. Initially they worked jointly on Mir, this co-operation being termed 'phase1' of the programme, with the ISS being phase 2. The Shuttle was to bring up supplies, equipment and astronauts. With their space-station experience going back to Skylab, the Americans were able to 'fill a gap' in their space programme.

The first illustration of this co-operation came about with the Spektr module. This was originally a military craft, designed as a response to the American 'Star Wars'

The Kvant-2 module docked on 6 December 1989. Although it bears the same name as the preceding module, it is quite different. It does carry scientific instruments, but it is chiefly an extension of the cosmonauts' living space with (at last) a shower worthy of the name and an oxygen and water recycling system.

Pictures of life aboard the stations, whether taken on Salyut, Mir or the ISS, reveal an incredible tangle of electric cables and straps holding notebooks, manuals or even teaspoons! In this apparent shambles, everything is actually in its ideal place and within handy reach. The crews need to be able to get at anything within seconds.

Krikalev is one of the most experienced of the former USSR's cosmonauts. After the TM-7 mission, he returned on TM-12 to be part of the ninth permanent expedition, but, for cost reasons, remained on board Mir when the tenth party arrived. The flight engineer would thus stay almost 312 days in space. This Soviet citizen returned to Earth a Russian, as in the meantime the Soviet Union had imploded. In 2005 he became the man who had spent the longest continuous time in space, beating the previous record of 745 days held by Sergei Avdeyev. He actually stayed aboard Mir for 151 days followed by another 312 days, before joining the ISS for stays of 141 days and 179 days.

Atlantis
US

programme. But the fall of the Soviet Union changed the game and it was decided to completely revise Spektr's equipment. The Americans agreed to finance part of the module, as long as some American-made equipment was used. But the different technologies used by the two countries delayed the project throughout the whole of 1994. The module finally docked with Mir on 20 June 1995. It included a number of scientific instruments for Earth observation and had an airlock and a robot arm to put small satellites into orbit. In addition, it served as a rest and living area for the American crew.

Co-operation between the two countries grew with the arrival of the Shuttle *Atlantis* on 29 June 1995 – a truly memorable event. Shuttle mission STS-71, with five crewmembers on board, was unique. For the first time, a Shuttle had docked with a station, 20 years after the historic Apollo-Soyuz flight. Mir was occupied by five other cosmonauts – including an American scientist. *Atlantis* had to get into position below the station to dock with Kristall's main port. Docking a Shuttle with Mir had been intended from the very start, but it would, of course, have been the Soviet Buran Shuttle, which was to have docked with the new Mir-2 station. The concept of a Shuttle-to-station docking module had not been abandoned, however, and in June 1995 such a module – developed jointly by the Russians and NASA – was delivered to the Kennedy Space Center to be taken up by *Atlantis*. The American Shuttle could dock directly with Mir without this module – which it did during the STS-71 mission – but Kristall's solar panels made the manoeuvre difficult. To obviate any risks, the docking module was installed on 12 November 1995, making future docking with the orbiters much easier.

Finally, on 23 April 1996, the last element needed to complete the job was Priroda, a Russian Earth-surveillance module originally designed for mixed civilian and military use but redesigned after the end of the old regime for a uniquely civilian application. Weighing almost 20 tons, it had its own thrusters for manoeuvring, but the solar panel providing its electricity had to be brought up by a Progress vessel and installed during an EVA. Living on board Mir was a remarkable experience. On this immense complex weighing over 100 tons, the greatest amount of space was devoted to scientific equipment and, while everything was strictly in its proper place, photographs reveal a tangle of cables and tubes.

Space goes international

With its seven modules, Mir resembled a labyrinth, a disconcerting impression reinforced by the fact that there was no up and no down, every corner and every wall being taken up by equipment. In a house, there is always a floor and a ceiling, but Mir's essentially cylindrical shape and the absence of gravity meant that such conventional arrangements were redundant.

Atlantis has just docked with the Mir station. It is 29 June 1995 and Mir is welcoming an American orbiter for the first time. It has brought two Russian cosmonauts, Anatoly Soloviev and Nikolai Budarin. The docking manoeuvre was carried out slowly, the Shuttle positioning itself below the station and approaching carefully. The manual process started at 2,600ft away, and at exactly 250ft it stopped to await the green light from the Russian and American flight directors. At a little under 30ft away, the commander, Gibson, began the final approach. At a speed of 1.3in/sec, *Atlantis* closes in to dock successfully with the Kristall module. The Shuttle's hold was fitted with an 'androgynous' Russian docking port, allowing it to dock with Kristall, as the special Shuttle docking module was not yet in place. When docked like this, the Shuttle and Mir formed the biggest and most imposing artificial satellite ever built.

The third joint Mir-Space Shuttle mission (STS-76) began on 24 March 1996. Seen here posing in *Atlantis*'s cockpit is the mission commander, Kevin P. Chilton, accompanied by Yuri Onufrienko.

Also on board the *Atlantis* was Shannon W. Lucid, the only American woman who had occasion not only to stay on Mir, for 188 days, but also to clock up a total of five orbital flights.

Co-operation between the countries, however, ran up against a number of difficulties, to the extent that NASA considered leaving the programme. But the new director, Daniel S. Goldin, put all his weight behind the notion of Mir being an international effort. Apart from the problems associated with language and different attitudes and viewpoints, the differences over what sort of experiments to conduct often gave rise to friction between those responsible.

Mir also experienced some serious incidents in 1997. On 23 February, the station was occupied by astronauts Vasily Tsibliev and Alexander Lazoutkin and a German specialist from the ESA, Reinhold Ewald. They had arrived on board Soyuz TM-25 on 12 February. The American Jerry Linenger had arrived a little earlier on the Shuttle *Atlantis*, and also on board was Mir's crew, Valeri Korzun and Alexander Kaleri. The latter were due to go back down with Soyuz TM-24 (accompanied by Ewald) at the beginning of March 1997, but they had been on board the station for nearly 190 days.

On the evening of 23 February, an oxygen

canister exploded and caught fire in Kvant-1. At the time, the crew was in the central module. A thick cloud of smoke began to fill the module and threatened to spread throughout the station. Worse still was the fire. Korzun and Kaleri grabbed fire extinguishers while the others tried to get rid of the toxic smoke. The fire itself lasted only 90 seconds, but for more than two and a half hours the astronauts had to wear gas masks, after which they took air, blood and urine samples for analysis. The water from the extinguishers had also increased the humidity in the station and the temperature had risen. Tired and somewhat shaken, the men faced yet further trouble, as the accident had reduced the oxygen regeneration. They tried to produce oxygen from their water reserves, but the amount needed was too great. In short, everyone was keen for the arrival of Progress M-34, bringing vital repair equipment for the oxygen plant. Various anomalies began to appear in the functioning of the station. At the beginning of April they had to cut out the carbon dioxide filters, which were overheating abnormally. Finally, on 8 April, Progress M-34 and its precious cargo docked. Using Progress's engine, the station was able to move into a higher orbit.

On 25 June 1997 a docking/undocking manoeuvre was due to take place to test the new auto-guidance system, which was replacing the old automatic system that was produced in the new state of Ukraine and the Russian Confederation could no longer afford to buy! Tsibliev began to get ready for the manoeuvre on his control screen. But the remotely controlled Progress deviated from its course and tore off one of Spektr's solar panels before colliding with it and piercing a hole. The accident quickly turned into a catastrophe, with the module depressurising rapidly. The whole crew could hear the hiss of escaping oxygen. Fortunately, they rushed to close the module's airlock, disconnecting all the cables. Disaster was just avoided, but once again the nerves of the American C. Michael Foale (who arrived on 15 May with *Atlantis* mission STS-84) and of the two Russians had been put to a severe test. Cut off, Spektr could no longer supply them with electricity and would later be condemned.

Tsibliev was clearly out of luck: this was the second time he had been in a collision with Mir – the Soyuz TM-17 vessel he was piloting in 1994 had also hit it, but had caused much less damage. Most of the repairs were carried out using EVAs, but it was becoming clear that Mir was ageing, like a rusty hulk. However, Mir's balance sheet looks impressive: 57 expeditions, 78 EVAs, 5,511 days in orbit (4,594 of which were manned) and a total of 98 astronauts on board. On 16 June 2000, Soyuz TM-30, with Kaleri and Sergei Zaliotin aboard, made a perfect landing back on Earth. They were the last to leave Mir, and on 23 March 2001 it was slowly taken out of orbit and numerous observers saw dozens of little shooting stars over the Fiji Islands as it broke up on re-entry. Nothing was left of it, but asteroid 11 881 has been named 'Mirstation' in its honour.

Linda M. Godwin is one of NASA's forgotten people, yet she flew four times, spent over 38 days in space and carried out a six-hour EVA during mission STS-76, the latter a first for an American on a Russian station. These kinds of operation are very much routine these days and everyone, including civilian specialists, knows how to cope with extra-vehicular activities.

Terence W. Wilcutt, *Atlantis*'s pilot, watches the Mir space station from flight STS-79. On this fourth mission involving the American orbiter and the Russian station, *Atlantis* was carrying a Spacehab with two modules, one intended for experiments to be conducted during the Shuttle's flight and the other to be transferred to the station.

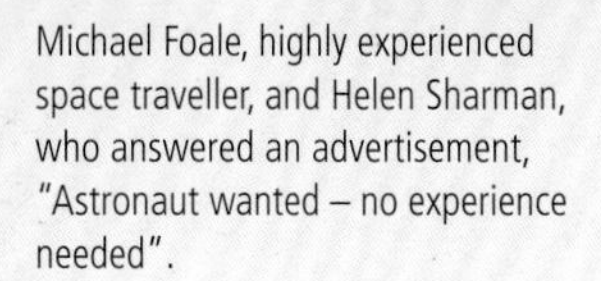

Michael Foale, highly experienced space traveller, and Helen Sharman, who answered an advertisement, "Astronaut wanted – no experience needed".

UK astronauts and Mir epic

Michael Foale, born on 6 January 1957 in Louth, graduated from Cambridge University in 1978 with a degree in Physics, Natural Sciences Tripos, with first-class honours. After completing a doctorate in Laboratory Astrophysics in 1982 he moved to Houston, Texas, to work on Space Shuttle navigation problems at the McDonnell Douglas Aircraft Corporation. At NASA Johnson Space Center, he was responsible for payload operations on Space Shuttle missions STS-51G, 51-I, 61-B and 61-C.

In 1987 Foale was selected as an astronaut candidate. Before his first flight he flew the Shuttle Avionics Integration Laboratory (SAIL) simulator to provide verification and testing of the Shuttle flight software, and later developed crew rescue and integrated operations for the International Space Station (ISS).

Foale has been a crew member on six space missions. STS-45 (1992) was the first of the ATLAS series of missions to study the atmosphere and solar interactions. STS-56 (1993) carried ATLAS-2 and the SPARTAN retrievable satellite that made observations of the solar corona. STS-63 (1995) was the first rendezvous with the Russian space station Mir. Foale made his first EVA with Bernard Harris for 4 hours 39 minutes. Their task was to evaluate extremely cold spacesuit conditions and explore mass handling of the 2,800lb Spartan satellite.

Selected for an extended mission aboard Mir, Foale was launched on STS-84 on 15 May 1997 to join the Mir 23 crew. When the space station was damaged after being struck by a Progress resupply ship, Foale conducted a six-hour EVA in a Russian Orlan spacesuit with Anatoli Soloviev to inspect the station's Spektr module. Also, drawing on an ancient technique, he was able to calculate how the stars were moving past his fixed-point thumb reference on a window and could thus advise Russian ground control how to stop the resulting roll. On 6 October 1997, after spending 145 days in space, Foale returned to Earth on STS-86.

In December 1999 he flew on STS-103, an eight-day mission, to repair and upgrade the Hubble Space Telescope.

On 18 October 2003 Foale launched on Soyuz TMA-3 and docked with the ISS, where he and Alexander Kalerie stayed until 29 April 2004, and conducted an EVA lasting nearly four hours. Foale served as Expedition-8 Commander. Mission duration was 194 days, 18 hours and 35 minutes.

Foale has logged over 374 days in space, including four space walks totalling 22 hours and 44 minutes.

Helen Sharman was selected as the British cosmonaut on the 1991 Soviet space mission Project Juno after replying to a radio advertisement: 'Astronaut wanted – no experience necessary'. As part of the Soyuz TM-12 mission she spent 7 days, 21 hours and 15 minutes on Mir and became the first non-Russian, non-American woman to go into space.

Born on 30 May 1963 in Sheffield, Sharman graduated from Sheffield University in 1984 with a BSc in chemistry, followed by a PhD from Birbeck College, London. Part of her job as a research technologist for Mars Confectionery was to study the chemical and physical properties of chocolate.

Sharman was selected for the Russian scientific space mission from more than 13,000 applicants. The only requirements were that they should be fit, British, aged between 21 and 40, with a science background and the ability to learn a foreign language. After weeks of exhaustive physical and psychological tests, she underwent 18 months of gruelling training at the Cosmonaut Training Centre in Moscow's Star City.

The Soyuz TM-12 space capsule was launched from the Baikonur cosmodrome in Kazakhstan on 18 May 1991. The mission lasted eight days, mainly spent at the Mir space station. Sharman's tasks included medical and agricultural tests, photographing the British Isles, and participating in an amateur radio hookup with British schoolchildren. She returned to Earth aboard Soyuz TM-11.

Sharman was just 27 years and 11 months old when she went into space and was, as of 2007, the fifth youngest of the 455 individuals (90% men) to have flown in space.

Soyuz T and Soyuz TM: the buses of space

The T and TM versions of the Soyuz vessels are modernised versions of the well-known vehicles used by the Soviets, and then the Russians, since the end of the 1960s. The Soyuz T appeared for the first time in 1978. Compared with the older craft it is equipped with a new Igla automated docking system, a service module with Vernier engines independent of the main engines, a greater fuel capacity and much improved avionics, and has a crew increased to three, all able to wear spacesuits. In addition the escape system is improved. The most recent development is the Soyuz TM, also designated '7K-STM' by its manufacturer. Intended for docking manoeuvres with the Mir and ISS stations, the Soyuz TM differs from Soyuz T by its descent module's lighter heat shield. It is also fitted with the more effective and lighter Kurs docking system. The Soyuz TM is the most frequently used vehicle for transporting crews up to orbital stations, as well as for evacuating them in the event of an emergency.

The International Space Station

The International Space Station (ISS) followed on from an American project initiated in 1984 under Ronald Reagan's presidency. The Freedom station was to have been the biggest ever built, significantly more modern than the current Salyut, and to have been permanently manned. But while NASA, under the leadership of James Montgomery Beggs (who stayed in the job until 1985), took up the challenge, it did not really know where to start the programme. Freedom's development costs very soon exceeded predictions. Like the other stations, Freedom was to serve not only as an Earth-observation post and low-gravity laboratory, but also as a 'garage' to repair damaged satellites or as a staging post for more-distant missions. Linked with the space Shuttle, Freedom was undoubtedly the programme NASA lacked; one that would have breathed new life into it, especially as the European Space Agency, Canada and Japan had wanted to participate in the project. But it went through no fewer than seven major revisions up to 1993. Besides these delays and internal dissension, the *Challenger* disaster had an immediate impact on the project. A commission was set up to look into all the plans in minute detail to assess their safety aspects. President Reagan gave way to Bill Clinton on 20 January 1993 and a weary Congress began to be much less inclined to fund this colossal project. Morale sank again.

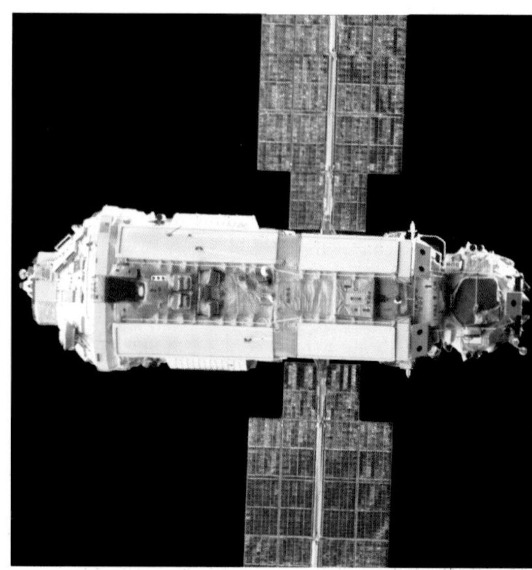

Freedom becomes the ISS

In 1993 the Clinton administration announced the end of Freedom, but resuscitated the station under the name of Alpha, including the Russians in the programme. Though less ambitious, Alpha was financially more realistic, Clinton having simply divided the budget by two! For their part, the Russians had gained substantial experience with Mir, but were faced with the same budgetary constraints as NASA. After the fall of the Soviet regime, the Mir-2 programme barely seemed to figure; likewise for the Buran Shuttle. However, the Mir-2 programme did not disappear completely, as the ISS's Sveda module is none other than the 'double' of Mir-2's DOS-7, the DOS-8 central module.

The Alpha programme started off with Mir when the United States and Russia signed an agreement allowing the Shuttle to make nine dockings with Mir, NASA alone providing the funding for arranging these flights. From 1997, Brazil joined the programme, which then became the ISS (the Russians had not been happy with the Alpha name). Just like Mir, the ISS was a Meccano set, but an international one! The station was modular and could be modified over time. Some of the modules had docking capability, and could be moved around, replaced by more up-to-date versions or enlarged. There was certainly a common desire to go forward, but not everything was so easy. Five partners, including a total of 16 countries, had to work together, with the inherent difficulties of language, different agendas and many other unexpected events. Thus, in April 1997, the Americans discovered that the main central module that was to be the first to go into orbit was completely empty of equipment and had no engines! They had

The Unity module was the first element designed entirely by the Americans. It was put into service by mission STS-88 in December 1998. This module (also known as Node 1) was indeed the node of the station, being equipped with six docking points for transfers from one module to another. To connect all the various cables to Zarya, the Shuttle astronauts had to do three EVAs. Here, James H. Newman is busy making connections.

The Zarya module was the first to go up into space, carried by a Proton rocket, on 20 November 1998. It was to serve as a kind of advance base and communications centre before these functions could be transferred to other modules. Also known as the Functional Cargo Block, it was built in Russia by the Moscow firm of Krunichev, using American funding. Weighing around 19 tons, the module had three docking ports (one at each end and a third on the side). Sixteen fuel tanks for the engines were attached to the sides of the module. It was intended to remain in this original configuration for just a few months, but delays in the construction of the Zvezda module left it to carry out its mission in this condition for two years.

During *Discovery*'s supply mission STS-96 on 27 May 1999, astronauts Tamara E. Jernigan and Daniel T. Barry are out in space installing various pieces of equipment.

to move heaven and earth to convince Russian President Boris Yeltsin that it was necessary to get a move on… This FCB (Functional Cargo Block, or FGB in Russian), christened Zarya ('sunrise'), formed the backbone of the ISS. It belonged to NASA, who had paid for it, but its design was entrusted to Chelomei's research department, the KHSC (the Krunichev Space Research Centre), which acted as a sub-contractor to Boeing, the project manager.

During the first months of the ISS's construction, Zarya had to house a crew, supply on-board electricity, serve as a storage and communications area and be capable of self-propulsion. The 20-ton module was sent up on a Proton rocket on 20 November 1998,

The Zvezda is one of the most important modules of the station. Its resemblance to Mir is not surprising, as it is the same module. Mir's three distinct parts can be seen: the sphere forming the transfer compartment with three docking ports, the work compartment, and a rear transfer section where Soyuz and Progress craft dock. This module received most of the missions originally destined for Zarya.

being joined on 4 December by the American Shuttle, *Endeavour*, which brought the module Unity (also called Node 1). This was an interconnection system for the other modules, with a total of six docking ports. Though smaller than Zarya, Unity weighed no less than 11.4 tons. Members of *Endeavour*'s crew (mission STS-88) on this occasion carried out three EVAs to connect all the electrical systems between Unity and Zarya. The astronauts even went inside one part of the module to check the communications equipment, but did not live in it.

Several other logistical missions, all undertaken by Shuttles, were intended to prepare the venue for its future residents. On 29 May 1999, mission STS-96 (*Discovery*) brought up nearly two tons of equipment from Spacehab and the ICC (Integrated Cargo Carrier) system in its hold. Tammy Jernigan and Dan Barry donned their spacesuits to put various items outside the station, and then, on 25 July, the third Sveda module docked with the station automatically, just behind Zarya. This was the former central module from Mir-2, modified to suit its new role. It represented one of the main elements for the Russians and, initially, it complemented Zarya as living quarters for the crew, providing a bit more comfort. It was equipped with the Elektron system, which was able to separate out (reused) oxygen and (rejected) hydrogen from the ambient humidity. Fitted with a shower and toilets, but also carrying the Russians' primary computer systems, in use the Sveda module turned out to be a particularly noisy and uncomfortable place! For budgetary reasons, NPO Energia did not have the wherewithal to make a back-up for this module, so it was absolutely vital that Sveda was not lost during the launch. Prudently, the Americans had constructed a rescue vehicle, the Interim Control Module (ICM), just in case... However, the ICM could act only as a propulsion system and could not shelter a crew.

The third logistical mission (STS-106), undertaken by the Shuttle *Atlantis*, took place

Far left: Attached to its external tank, *Atlantis* takes off on 7 February 2001. On this mission, STS-98, the Shuttle will stay in space for more than 12 days, beating its previous record. In its hold it carries the Destiny module, an American laboratory that will dock with the international station.

Left: *Atlantis* takes off on 8 September 2000 on the third mission to the international station (STS-106). This was another supply and maintenance mission, but the aim was to prepare the station for the first expedition.

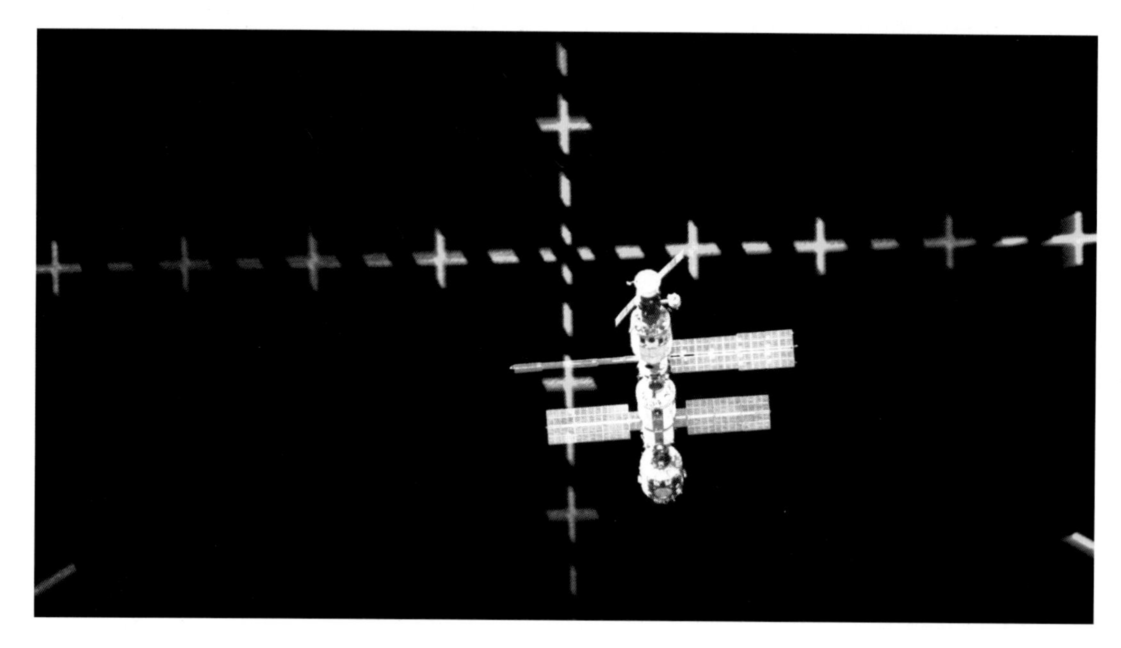

The station as it appeared on the final approach (top) and after completion of the work (below). The three modules, Unity, Zarya and Zvezda, are joined. The Soyuz docked at the end is in fact the Progress M1-3 that took off on 6 August and docked two days later. It was carrying supplies and equipment that the *Atlantis* astronauts transferred onto the station.

in September 2000. At this point the two modules were connected to each other. *Discovery* carried out one more mission on 11 October prior to the station's occupation – it was the 100th Shuttle mission. The crew fixed the pressurised Z1 structure to the Unity module, which then allowed them to attach the long trusses supporting the solar panels. Gyroscopes and communication and temperature-control devices were fixed to the structure. Z1 was the first component of the ITS (Integrated Truss Structure), which was to form a large non-pressurised assembly carrying numerous items of equipment. It was also to serve as the rail for the Canadarm 2 robotic arm when this was installed on its automated platform.

On 2 November 2000, the first expedition to

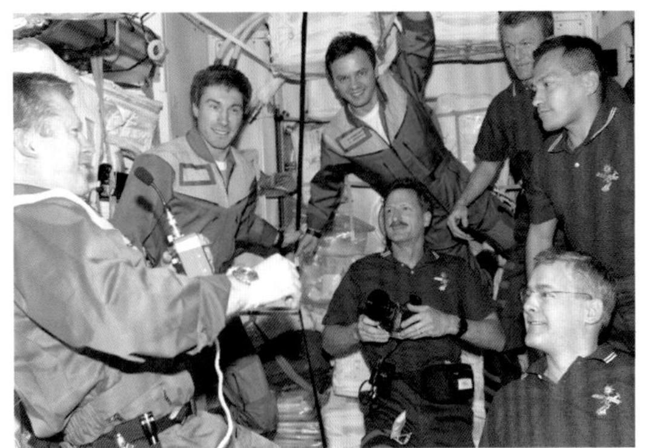

The Shuttle's robot arm placed the Destiny module's 14 tons of aluminium in position. *Atlantis*'s crew then made a number of EVAs to complete the final assembly.

Far left: Some crewmembers lend a hand in preparing the Destiny module. The internal equipment is configured around payload racks, which allows greater flexibility. The module is self-sufficient, with its own electricity supply, cooling-water circuit and air regeneration system. Over a period of time more racks have been added.

Left: After mission STS-92 in October 2000 (installation of gyroscopes and the large truss for the solar panels), the November mission (STS-97) assembled the panels and made preparations for the arrival of the American Destiny module. On board, the orbiter's crew met the ISS's first residents, who had come up on 31 October on Soyuz TM-31. Seen here are (in blue, left to right) William Shepherd, Sergei Krikalev and Yuri Gidzenko, chatting to some of STS-97's crew (in red).

the ISS set out. It comprised an American, William Shepherd, and two Russians, Sergei Krikalev and Yuri Gidzenko. They took off from Baikonur in Soyuz TM-31 on 31 October and reached the ISS on 2 November. On 17 November, after its initial approach failed, they were supplied by the Progress M1-4 ship. A few days later they were joined by the Shuttle *Endeavour* (mission STS-97), carrying in its hold four large solar panels and some heat dissipaters, as well as the arm (type P6) on which they were to be supported. These solar panels were to provide most of the on-board electricity and required lengthy EVAs to install them.

The Shuttle *Atlantis* (mission STS-98) brought up the Destiny module on 7 February 2001, the first proper American scientific module since Skylab. Designed by Boeing, this 16-ton, cylindrical, aluminium module permitted a range of scientific experiments using a multitude of removable racks. New racks were regularly brought up on missions by other Shuttles. The second crew to take up residence in the ISS arrived on 10 March 2001 on board Shuttle mission STS-102. Comprising Yuri Usachev (Russian Confederation), Susan Helms, (United States) and James Voss (United States), it relieved the previous crew before handing over to the third expedition.

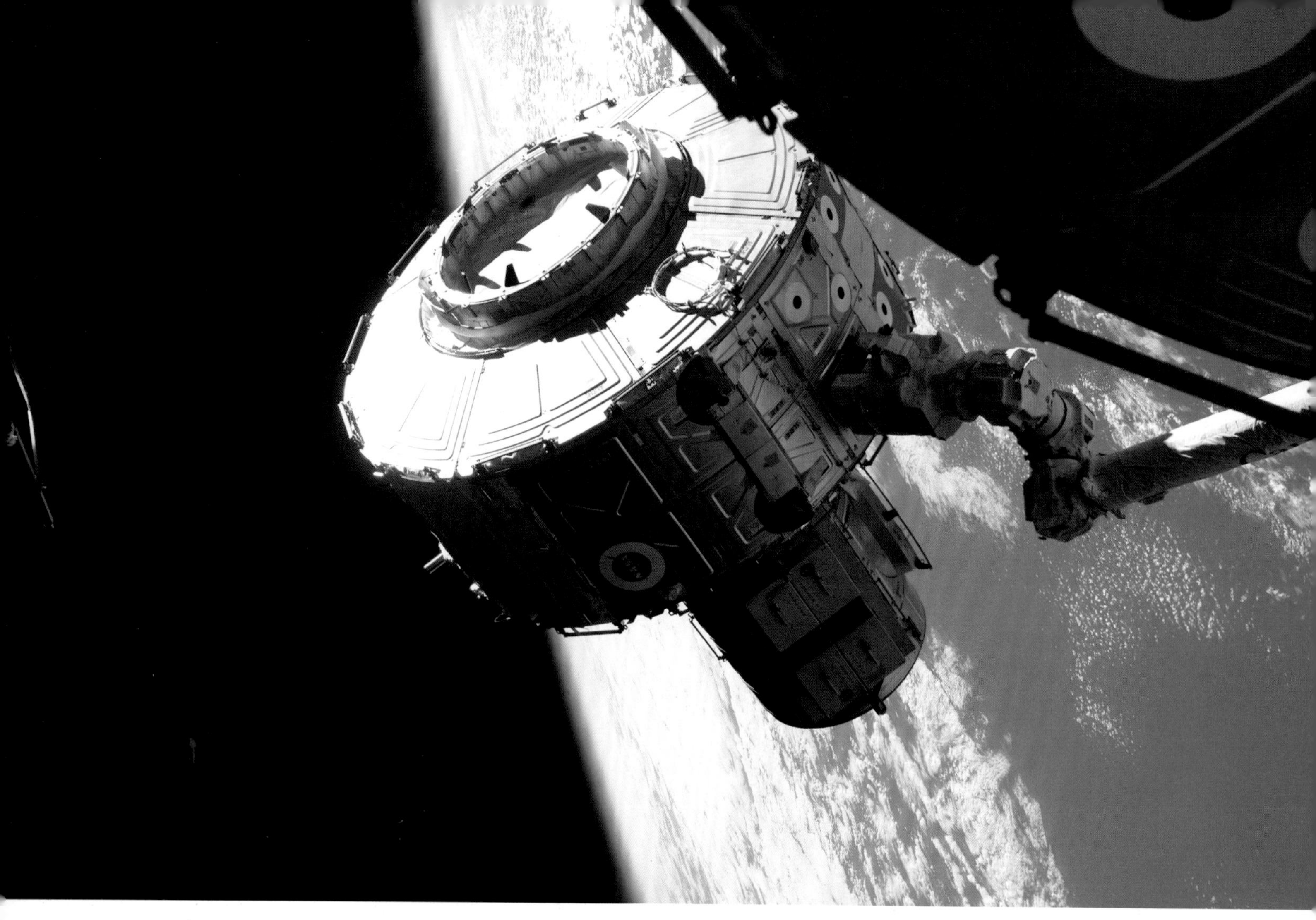

After the relief of the ISS's permanent crew, construction proceeded with the installation of the Canadarm 2 robotic arm, the Italian Raffaello module and the airlock. This decompression module has room for two astronauts to leave and re-enter the station, whether they are wearing an American EMU spacesuit or the Russian Orlan model. Made by Boeing for the Marshall Space Flight Center, the airlock is 20ft long and has a diameter of 13.1ft. Seen here (opposite page, top) is James Reilly, one of the members of mission STS-104, leaving the module to go out into space.

In the meantime the station had had grown further with the addition of a Canadarm robotic arm brought up by *Endeavour* (mission STS-100) on 19 April and, on 12 July, the Quest module – a universal port allowing astronauts wearing either an American or Russian spacesuit to pass through. The third permanent crew took over on 10 August, the same date that the Russian SO1 Pirs module was attached. This docking module was almost identical to one of those used on Mir. It allowed docking with Soyuz TM or Progress craft, and also made it easier for Russian astronauts wearing Orlan-M pressurised suits to carry out EVAs. It was launched by a Proton rocket before being propelled to the station by a Progress craft. Right at the end of 2001, the fourth permanent crew was brought up on *Endeavour*, ready to get down to work.

In spring 2002 various new structures were sent up and attached to the station along the trusses. These structures were built around a central truss (SO) mounted above the laboratory module, Destiny. There was no particular order in the way these elements were brought up. Thus, while the SO was the central master truss, other trusses (such as P6)

Yuri Usachev is one of the three cosmonauts from the second expedition to the ISS. He arrived aboard the Shuttle on mission STS-102 on 9 March 2001, and returned to Earth 163 days later. He is caught here in the Zvezda module preparing food. On the left, hanging on the wall in packets, are pre-impregnated washing-up cloths, and just below them sachets of mayonnaise and ketchup!

had been sent up earlier, but it was a simple matter to remove these pieces and re-install them as the ITS structure grew. Furthermore, nothing was set in stone. With SO being the main truss, those located to the right of it were identified by the letter S (starboard) and those to left by the letter P (port). Unhelpfully, the piece identified as Z1 was so named because it was attached to port Z on the Unity module – but it was still part of the Integrated Truss Structure.

Mission STS-110 included the MT, or Mobile Transporter, a sort of wagon on castors, which ran on rails formed by the trusses later used for the Canadarm 2 robotic arm. This indispensable device could move almost anywhere the length of the station. *Endeavour* took off in June 2002 with the fifth permanent crew and, in its hold, the MSS or Mobile Servicing System, which allowed the previously delivered MT's capabilities to be fully exploited. Before 2002 was out Shuttle flights STS-112 and 113 delivered elements S1 and P1 of the ITS. With STS-113, expedition number 6 took over from number 5, which was the last occasion for some time that a crew was able to return to Earth aboard a Shuttle.

There were numerous EVAs during mission STS-110 to fix the central truss section supporting the remaining trusses that would eventually take the huge supplementary solar panels and other equipment (radiators etc). This central section was designated the SO or Center Integrated Truss Assembly Starboard Truss. It had to serve not only as a support structure, but also to conduct the power lines, cooling-water pipes etc from other modules.

The International Space Station in 2002.

Columbia fails to respond

While the space Shuttles were heavily used on flights to the ISS, NASA also wanted to run its own missions. *Columbia*'s STS-109 on 1 March 2002 was the first mission for two years that was not conveying equipment to the ISS. Its purpose was to carry out some routine maintenance on the Hubble telescope, after which flights supplying the ISS would start again. On 16 January 2003, on mission STS-107, *Columbia* took off on its last flight. It was the oldest of NASA's orbiters and it was also the only one not to have been brought up to standard for the most recent flights. Heavier than the others, it could make only what were known as 'solo' orbital missions, as it lacked the modifications required to link up with the ISS. *Columbia* was also the only orbiter in the fleet to have kept the TPS heat-resistant tiles – but during the course of its career it had been given heat protection on its upper surfaces like the other Shuttles.

Mission STS-107 was the first to test Spacehab. This module was the first to have been designed and offered to a space agency by a private company. The pressurised module was carried in the Shuttle's hold along with other equipment, depending on the mission. STS-107 was to carry out experiments that could be done on the ISS, but by a smaller crew.

Under pressure from Congress, which was also keen to see missions that were 100 per cent American, *Columbia*'s flight was officially set for the beginning of 2003, after many postponements – the flight had originally been scheduled for 2001 – which explains why it took place after missions STS-108 to STS-113. *Columbia* took off normally and entered orbit for a mission of over two weeks. On board, the mission commander Rick D. Husband and pilot William C. McCool were accompanied by specialists David M. Brown, Kalpana Chawla, Laurel B. Clark, Ilan Ramon (the first Israeli astronaut) and Michael P. Anderson, a retired pilot. The last two were in charge of supervising operations in the hold.

Having completed its mission, on 1 February the doyen of Shuttles positioned itself, nose up, for re-entry at an altitude of around 75 miles. It was at this always tricky stage that the problems began. The deceleration stage naturally caused *Columbia*'s underside to heat up, but the sensors on the left wing were indicating high temperatures in the area of the undercarriage and flaps. Then, at 14:53 GMT, the ground controllers noticed that the heat sensors had stopped working around the hydraulic systems, also on the left side. But as this had already occurred without the crew reporting anything abnormal, nobody worried unduly about it.

Three minutes later, the temperature had risen in the undercarriage well and the tyre pressure was increasing. This time it was abnormally high. This had to mean that the thermal protection was not doing its job properly and the wing's aluminium structure was starting to melt. The first piece of debris came off, a thermal tile that was found in New Mexico. An alarm should have sounded in the cockpit at this point. At 14:58 several other sensors stopped transmitting. One minute later, Houston asked the mission commander

The crew of STS-107 gets ready to board *Columbia*. Left to right are (front row) Rick Husband and William McCool; (second row) Kalpana Chawla and Laurel Clark; (third row) Ilan Ramon and David Brown; and right at the back Michael Anderson. Ilan Ramon was the first Israeli to go into orbit.

The remains of *Columbia* – around 2,000 pieces – were scattered over the United States between eastern Texas and Louisiana. The authorities immediately warned people that some of the debris might be dangerous. The remains were later taken to the Johnson Space Center in order to reconstruct as much of the Shuttle as possible.

to take a look at his undercarriage tyre-pressure gauge. At Mach 18 and an altitude of 40 miles, Husband barely had time to reply. Transmission ceased and *Columbia* broke up over Texas and Louisiana. Thousands of pieces of debris that had not been burnt up were scattered over an area of almost 2,000 square miles and were recovered and laid out at the Kennedy Space Center. There were no survivors.

It had all started on lift-off, less than ten seconds after ignition. A piece of insulating foam from the bipod linking the Shuttle and the external tank had come away and hit the leading edge of the Shuttle's left wing and damaged it (even though it was carbon-fibre reinforced). It had thus created a weakness just at the spot where the heat is most intense during re-entry.

Once more, all the Shuttles were grounded. The *Columbia* Accident Investigation Board had to analyse 84,000 pieces from the Shuttle. Initially it seemed improbable that a mere piece of insulating foam could have destroyed the Shuttle. Indeed, many cases of foam coming away from various parts of the external tank had already occurred without any effect on the missions. However, several tests demonstrated that a piece of foam hitting the leading edge of the wing at speed could seriously damage it. But that was not all. The commission of enquiry also pointed the finger at certain working procedures in the American administration that had placed too much confidence in ageing equipment. In the meantime, the ISS continued to need supplying. Only the Soyuz craft were available in sufficient numbers, but the whole programme now needed to be looked at again.

The loss of *Columbia* brought all Shuttle flights to a halt and no further modules could be sent up. In the meantime, the rotation of permanent crews was covered by the Russians, but the capacity of each Soyuz launch was restricted to two men. Here expedition 7 takes off on 26 April 2003 with the Russian Yuri Malenchenko and the American Edward Lu.

The Russians to the rescue

The ISS's expedition number 6 had just been re-supplied by a Progress ship when its crew learnt of the tragic end of *Columbia* and its seven occupants. Another Progress arrived on the 4th, but while the Russians were able to provide the supply logistics, the provision of new modules was NASA's responsibility. The construction of the ISS was consequently subject to considerable delay, with American orbital flights interrupted for two and a half years. NASA contributed the most financially to the project, and however much the Russians might justifiably be proud of their Proton rockets and Soyuz and Progress spacecraft, they could not on their own manage to launch new modules. Moreover, from expedition number 7 onwards (April 2003) it was decided to send up only two astronauts at a time using a Soyuz TMA ('Transport Modified Anthropometric').

The missions had to be rapidly revised and the programmes were simplified. The station needed to be maintained for a period of time,

Edward Lu in the Destiny module. He remained in the station with Malenchenko for 185 days. The Russian even got married by live link, taking his vows more than 200 miles above the Earth!

The TMA vessel was one of Soyuz's final variants, the fifth to be exact. It was specially modified to dock with the ISS station, and had various improvements that provided a greater level of comfort than the old TMs.

but all the partners recognised that the Shuttle's absence was highly prejudicial to the whole of the ISS programme. Without the Shuttle, it was difficult to carry out EVAs – human intervention being indispensable to assemble and connect the modules. With a Progress vessel taking a much smaller payload than a Shuttle's hold, rubbish began to accumulate. In short, if the number of Russian launches was going to have to be increased, the ISS's financial viability would become an issue.

It was not until 26 July 2005 that the Shuttle *Discovery* brought this standstill to an end. This mission (STS-114) put the External Storage Platform (ESP-2) into service. But once again, despite all the precautions taken, a piece of foam came away from the Shuttle. It caused no damage, but NASA once again had to suspend flights. Consequently mission STS-115, taken up by *Atlantis*, could not take off until 9 September 2006, when it brought the station new trusses to support solar panels. By this time the programme was almost four years behind schedule – the other modules would have to be dispatched bit by bit up until 2010, by which time some of the station's older parts will be reaching the end of their useful lives and will need to be replaced. Nevertheless, the ISS has been, on balance, a success. By June 2008, 163 astronauts from many countries had visited it at least once, and five of them had been 'tourists' (the Americans Dennis Tito and Gregory Olsen, the South African Mark Shuttleworth, the American-Iranian Anousheh Ansari and the Hungarian-American Charles Simonyi), who paid some $20 million each for their space tickets. A wedding was even celebrated live from the ISS on 10 August 2003, between the Russians Yuri Ivanovich Malenchenko and Ekaterina Dmitrieva.

The American orbiters made their comeback with *Discovery* joining the ISS on 28 July 2005. New safety measures had been worked out for this and all ensuing flights. The robot arm was fitted with a new Orbiter Boom Sensor System (OBSS) that allowed it to check the whole of the heat shield and detect any dislodged tiles. In fact, this is exactly what had happened on take-off, when one of the tiles was torn off before the separation of the external tank.

All hands on deck! Some of *Discovery*'s crew on the Shuttle's upper deck watch the gentle approach to the station.

NASA's Orion project is very similar to Apollo. The command module's design, put forward by Lockheed-Martin, is virtually identical. Originally christened the 'Crew Exploration Vehicle' (CEV), it is likely to replace the orbiter in around 2010. Orion will be launched by a new Ares rocket (former CLV, Crew Launch Vehicle) designed by Thiokol, the manufacturer of the Shuttle's boosters. Unlike Saturn, Ares has a solid-fuel first stage, with the second stage being a development of the J-2 (J-2X) engine that propelled Saturn. A more-powerful version, Ares V, has also been proposed to send much heavier loads (up to 130 tons) into low orbit. This will be a fairly standard rocket, with five RS-68 engines (like those on the Delta IV) between two separable boosters of the same type as the Shuttle's but more powerful.

China's awakening...

China began to study the possibility of sending a man into space quite early. An initial project, christened 'Shunguang-1' had been launched at the start of the 1970s, but it was abandoned. In fact the country was secretly planning a manned space programme worthy of the name throughout the 1970s and 1980s, but nothing seemed to come of it. It was only in about 1985 that China decided to get back into the space race with an initial, ambitious, Space Shuttle programme. However, Chinese aeronautical technology did not yet have the means to develop a hypersonic vehicle and the Shuttle was soon written off. To save time, the experts at the Chinese Academy for Space Technology eventually decided to adopt a design very similar to the Russian Soyuz, which could be built quickly and at reasonable cost.

After much equivocation, the Chinese programme number 921/1 was given the go-ahead in 1992. The craft was to be launched by a Long March rocket. Thanks to Russian co-operation, Shenzhou differed very little from Soyuz, replicating its three-module concept (orbital, re-entry and service). It was, however, a little bigger and, in particular, the orbital module was autonomous, having its own engines and its own control system, which suggests that at some point in the future each module might be able to dock with a proper space station. Overseen by engineers Qi Faren and Wang Yongzhi, the first unmanned Chinese spacecraft was launched on 19 November 1999. It was followed by three others before Shenzhou 5 took up the taikonaut Yang Liwei on 15 October 2003. The two-stage Long March 2F was fired from its launch pad in the Gobi desert and placed Shenzhou 5 into an orbit with an apogee of 208 miles. The craft and its pilot remained in space for over 21 hours. It then returned to Earth without any problems while the orbital module continued with its experiments until 16 March 2004, before burning up in the atmosphere on the 30th of the same month. The second Chinese manned flight took place on 12 October 2005. Shenzhou 6 was occupied this time by two men, Fei Junlong and Nie Haisheng. They tested the orbital module during the five days of the mission, much appreciating its comfort, and rejoined the re-entry module on 16 October (the orbital module eventually ceased functioning on 15 April 2006). Again, the return took place without any problems. The arrival of China in the family of nations that have sent astronauts into space did not evoke any great reaction, although most countries hailed the Chinese exploit.

Towards a new project?

On 14 January 2004, President Bush announced in a speech that the USA was to give NASA a new *raison d'être*, using the moon as a staging post for travel to worlds beyond our own. George W. Bush had just given the green light to a new project named Constellation. This new vision of space exploration strives to consolidate NASA's presence in Earth orbit, then to return to the moon around the year 2018, creating a base there before preparing new missions to Mars. This huge programme revolves around, among others, the CEV (Crew Exploration Vehicle), which will replace the Shuttle around 2010. Capable of travelling long distances outside Earth orbit, the CEV looks rather like an Apollo craft capable of docking with the ISS. Despite this design from the past, the CEV will naturally benefit from all the latest technological advances. It will also need a new heavy launcher of a lower capacity than Saturn (100 tons as opposed to 140 tons), but proportionally less expensive.

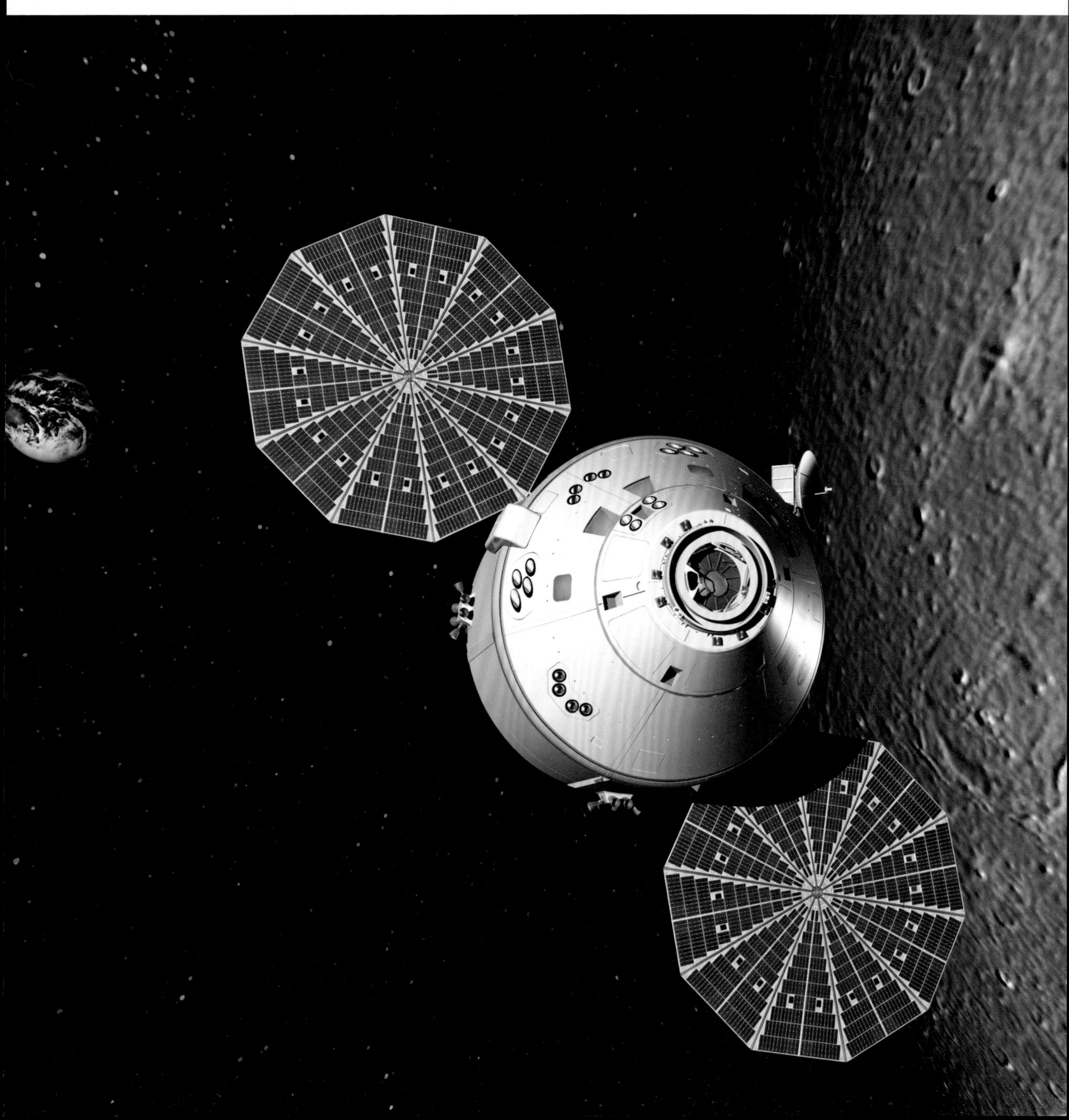

An artist's impression of Orion in orbit around the moon. The aim of the American programme is to return to the moon, setting up a base there, which will be used as a staging post for the exploration of other planets. Unlike Apollo, the Orion capsule would return to Earth, not splash down in the sea, and land with the aid of large parachutes and airbags. This method of operation should be substantially cheaper. In the same way, the command module could be reused about ten times. In short, the aim is to combine the advantages of Apollo and the Space Shuttle without their disadvantages.

INDEX